W9-AAD-958

GOD'S LAST OFFER

GOD'S LAST OFFER

Negotiating for a Sustainable Future

Ed Ayres

FOUR WALLS EIGHT WINDOWS
New York / London

Published in the United States by
FOUR WALLS EIGHT WINDOWS
39 WEST 14TH STREET, ROOM 503
NEW YORK, NY 10011
http://www.fourwallseightwindows.com

U.K. offices:
FOUR WALLS EIGHT WINDOWS / TURNAROUND
UNIT 3, OLYMPIA TRADING ESTATE
COBURG ROAD, WOOD GREEN
LONDON N22 6TZ ENGLAND

LIBRARY OF CONGRESS CATALOGUING-IN-PUBLICATION DATA

Ayres, Ed.
God's last offer : negotiating for a sustainable future / by Ed Ayres.
p. cm.
Includes bibliographic references and index.
ISBN 1-56858-125-4
1. Human ecology. 2. Environmental degradation.
3. Conservation of natural resources. 4. Sustainable development.
I. Title.
GF21.A93 1999
33.7'2—dc21 99-17049
CIP

10 9 8 7 6 5 4 3 2 1
Printed in the United States

Interior design by Terry Bain

CONTENTS

A DIMINISHING CAPACITY FOR ASTONISHMENT

THE WORLD we thought we knew has become a strange and agitated place in the past few years, and the change—like that of an ailing parent slipping into cantankerous dementia—has caught us off guard. What we have always assumed would be here to sustain and gratify us, give us purpose, forgive our mistakes, and encourage our dreams, we find suddenly failing. We ache for reassurance. But on three critical fronts, things have gone wildly out of control:

The global *economy*, celebrating the triumph of free-market democracy as the final decade of the twentieth century got underway—and anticipating the dawn of a new century of even more spectacular technological development and economic growth as the decade approached its end—unaccountably began staggering as though it had been poisoned. In 1998 came the cascade of Asian financial collapses—Indonesia's middle class thrown back into deep poverty; some of Japan's most vaunted banks closing their doors; and the gap between rich and poor growing rapidly worse in almost every country. Russia, which was supposed to be the greatest prize of the free-market victory, watched in agony as its currency plunged and its grain production fell to less than half what it had been just eight years earlier. It wasn't supposed to be like this.

The economic collapses came toward the end of a decade of staggering shocks to the global *environment:* vast fires raging out of control in Brazil, Indonesia, Malaysia, and Mexico; catastrophic floods in China, Bangladesh, and more than 50 other countries; an anomalous acceleration in global warming; the appearance of pesticide-proof insects that are eating as large a share of the world's food crops now as they did in medieval times; the worldwide resurgence of infectious diseases once thought to be conquered; and outbreaks of new diseases for which no cures are known.

These disruptions to money, land, and health were accompanied by an escalation of new, often inexplicable, *human conflicts and breakdowns:* the appearance of terrorists who are citizens of no country launching assaults that no military force knows how to defend against; the committing of genocides not by bitter ideological enemies but by close neighbors; the conscripting of children into guerrilla armies; the massacres of refugees; the increasing incidence of sociopathic behavior in almost every country; and the increasing incidence of *self-destructive* behavior.

If these trends are analyzed as separate sets of problems, it is hard to make any sense of them. They discourage and overwhelm us—and cause many to turn away and not want to think about them at all. Commentators and legislators analyze the latest news of them, while bureaucrats and consultants agonize over particular aspects of them; policies are proposed and hundreds of new laws and regulations are passed. But nothing seems to slow their overall momentum.

We are beginning to grasp that only by approaching these separate economic, environmental, and societal megatrends as being *fully interdependent*—no less connected than a bird's wings are to its circulatory system, its beak, and its brain—can they be explained in a way that will arrest them before our fall has become unstoppable. To understand what happened to the economy, for example, is impossible if we analyze it only in terms of interest rates, inflation rates, employment rates, currency exchange rates, and so on, as business analysts habitually do. It is also essential to take into account the changes in climate, forest cover, water resources, and biological processes that form an economy's real-world foundation. And it is equally essential to take into account the societal conditions that drive that economy: how many hundreds of millions of people still live on less than a dollar a day; how many now live in the growing shadow economy; how many have been driven from their homes; and how many are sick.

Perception of the world is not unlike the perception of a photograph of a human face, as printed in a newspaper or magazine. The photo is made up of dots, or "pixels," and a small piece of the photo seen close-up is unrecognizable. In today's high-speed world, each of us receives vastly larger numbers of "bits" of information about our world than earlier generations ever did, but those bits are still like the dots in an extremely tiny fragment of an increasingly enormous picture. From where we normally see it, it is incomprehensible. But stand back far enough, and the larger picture comes into focus. The world's multiple declines become visible as a single decline. It becomes clear that we are in a megacrisis of our own making, and that we have a chance now to escape it before it destroys us—but that the chance won't last long. The window of opportunity is closing fast.

How to see through that window before it's too late? A clue can be found in the ways human societies have long dealt with the things most critical to their present survival and future hopes: they tell eye-opening stories, whether in the form of myths, legends, or songs. Many societies, for example, have passed down stories of great floods. The Chewong people of Malaysia, the Koyukon of Alaska, the Maya of Mesoamerica, and the Christians and Jews who spread out from the Middle East all have told of great inundations. Whether those accounts contain dim memories of some prehistoric event that actually occurred or some intuitive grasp of what we can bring upon ourselves if we disrespect the powers of the world into which we were born, is hard to know. But that these stories offer something more than mere entertainment, I think few would deny.

Such stories may open eyes or minds by telling of events that were not anticipated—astonishing ordeals that might now serve as insights or warnings. For a story to be truly eye-opening, however, requires that much of the setting in which it takes place—the geography, climate, or culture—be familiar. There is a continuum of rising intensity between a light rain and a catastrophic flood, for example; if you have experienced the rain, you can mentally extrapolate to the flood. But if you have lived all your life in a place where no water has ever fallen from the sky and no sudden rivulet has ever run across the ground before you, you might find the idea of a terrible flood impossible to grasp. There is a paradox of perception here: that where there are no familiar conditions, there may be no galvanizing shocks. In a time of increasing disturbance and discontinuity, that paradox poses a growing threat to our ability to plan.

The records of history, as well as of recent psychological research, suggest that on those extraordinary occasions when people are suddenly confronted with something that is utterly alien to their experience, they may in effect go *blank* while the neurons race around in search of a familiar pattern of synapses —some memory, or myth, or clear expectation. Consciousness is a connecting of sensory stimuli and meaning, and if no connection is made, there may be a failure of consciousness. You may not see anything at all.

An incident that occurred more than two centuries ago, but was carefully documented at the time, illustrates what can happen. For millennia, the Aborigines of eastern Australia used small bark canoes to fish off the coast of their isolated continent. They had no contact with whites. But on April 29, 1792, the British sailing ship *Endeavour*, under captain James Cook, sailed into a bay and encountered a group of natives—the first known contact between Australians and Europeans. One of the passengers on the ship was an avid botanist, Joseph Banks, who was keeping a detailed journal of everything he saw on the journey.

To the natives, the sudden appearance of this ship would have been as unprecedented and inexplicable as it might have been for today's New Yorkers to look up and see a city-sized space ship blocking out the sky. The English ship, as described by historian Robert Hughes, was "an object so huge, complex, and unfamiliar as to defy the natives' understanding." In retrospect, it seems unlikely there was anything in their experience even to make them see it as a boat. In any case, in their reactions, they showed no evidence of seeing anything at all. The ship floated past the canoes, but as Hughes writes—based on Banks's account—"the Australians took no notice. They displayed neither fear nor interest and went on fishing."

We can speculate about what went on in the Aborignes' minds: that perhaps they were like the crowd in the tale of the Emperor's New Clothes, in which each person who saw a naked emperor simply thought that this could not be and was therefore unwilling to mention what he saw lest he be judged insane. But even in that tale, at least the idea of being naked is a concept everyone understands. But the idea of an enormous . . . *thing?*

In any case, the Europeans on the ship, having seen that no hostile responses had been provoked, lowered their landing boats and began rowing in to shore. As Hughes explains, what happened then was that the natives suddenly recognized something that *did* have meaning in their experience: "The sight of men in a small boat was comprehensible to them; it meant invasion. Most of the Aborigines fled into the trees, but two naked warriors stood their ground and shouted. . . ."

The 6 billion natives of Earth are in a position much like that of the Aborigines of the *Endeavor* encounter. We are being confronted by something so completely outside our collective experience that we don't really see it, even when the evidence is overwhelming. For us, that "something" is a blitz of enormous biological and physical alterations in the world that has been sustaining us. As happened with the Aborigines, it is an advent that will change life for us in ways we cannot conceive of. But there's at least one all-important difference between the situations. In the world at the outset of the third millennium, while there may be billions of people who don't see the thing that confronts us, there are at least a small number who have some inkling of it. They include our leading scientists and global-trends analysts, and they've been trying to arouse the attention of the rest of us—and they have failed. It is as though the wise elders of the Aborigines were standing on the shore waving their arms and shouting to the men in the canoes—but the men could not, or would not, hear.

The time has come for us to wrench ourselves from the stupor with which we humans have half-heard, or shrugged off, or blanked out, those warnings. We know, from our history as a species, that there are individuals among us who have great capabilities to respond heroically to challenges or threats that may seem overwhelming. But now we have come to a point where the courage of individuals won't suffice; we now need humanity *as a whole* to become heroic. We need, urgently, to pursue a question even more fateful to us now than whatever it was the European explorers were pursuing at that historic moment when Captain Cook was discovered by Australia: Why *are* we so unresponsive to the challenge that now looms before us? Why is it that even many of the most educated, news-watching, world-savvy people among us have only the vaguest awareness of the fact that the most world-changing event in the history of our species—more world-changing than World War II, or the advent of the nuclear age, or the computer revolution—is happening right now? What is going on to so profoundly block our perceptions of the fact that, so to speak, our ship has come in?

1

THE FOUR SPIKES

FOUR revolutionary changes are sweeping the world, and they will transform everything. Though these changes may be as momentous as our long-forgotten transition from hunter-gatherers to city-dwellers, or from animal-powered work to machines, most of us may be only vaguely aware that they are happening at all.

Of course, everyone who has dared to notice knows that we're in a time of growing turbulence: not a day goes by when there isn't more news of potentially world-changing consequence. Things are changing far faster than we can keep track of them, and there are some familiar catchphrases about how we react: we're "stressed out"; we're "overwhelmed." In the 1960s, car ads reveled in change: each new year brought dramatic unveilings of new models loaded with new features. Now, in stark contrast, a TV commercial depicts a woman being battered by the complications and stresses of everyday life—then offers an antidote. The noise stops, her frantic motion stops, and there before her is a car. Superimposed over the image of the car are the words "Simplify. Honda Accord." The ad is highly deceptive, of course. The internal combustion-powered automobile, the extended rush hours, the poisonous air that now shrouds Mexico City, Denver, and Beijing, and the global trade that drives the production of such cars ever upward, are big parts of what is overwhelming us. But the advertisers understand our stress. They also understand our confusion, and know we probably won't grasp the deep irony of such an ad—if, indeed, they grasp it themselves.

There are too many things happening for most of us to make sense of it all, and many people have given up trying. It seems, sometimes, as though we've been caught in a fire storm, with flak flying in all directions, and with no way of knowing what's happening or what is causing it. But if we can put this fire storm on replay, and track back through the webs of cause and effect, we can see a comprehensible—and startling—picture begin to emerge. On a graph of evolutionary change, four megaphenomena appear. Over a period of a thousand centuries or so, these phenomena look fairly stable—their rates of change as slow as the shifts of continents or the evolution of species. Then, within the span of just a few *years*—the years in which *we* live—they spike. The most disturbing of the changes we are experiencing—and will be experiencing much more profoundly in the next few years—can be traced largely to the effects of these four spikes.

In some respects, these megaphenomena represent momentous new opportunities for humankind—opportunities that may be sensed instinctively by people like the entrepreneurs of new technology who have made large fortunes in recent years. More demonstrably, though, they represent dangers of a magnitude that is hard to convey without seeming to lapse into hyperbole. The weight of scientific evidence now makes it clear that what we do now to confront and meet the challenge of these megaphenomena, will largely determine whether human civilization can survive in the long term—and whether our own generation will meet its rising expectations or enter a time of deepening impoverishment and regret.

The four spikes are causally connected, each adding fuel to the others. Together, they are hurling us into a spiraling loss of capacity either to see or to control where we are going. But before we can understand how these four phenomena interact, it's essential first to get a clear view of each of them individually, as a discrete trend.

THE CARBON GAS SPIKE

Everyone in the industrialized world has heard, ad nauseam, about the problem of global warming, or of the rise in carbon dioxide emissions that contributes to it. But there is little in what most have heard to suggest that this problem—or the endless wrangling over it by politicians and industries—is any different than the hundreds of other global issues that come and go in the news. Whether these issues concern global stock markets, illicit drugs, ethnic conflict, emboldened terrorists, military alliances, or the endangered environment, all seem to generate endlessly irreconcilable debate. So, we frown or shrug and move on to the next issue.

In 1997, global warming took its turn in the media (not for the first time) with the news that an international conference would be held that December in Kyoto, Japan, to negotiate a global climate treaty. The gist of the story was that some climate experts were saying the high levels of carbon dioxide (CO_2) pumped into the air was warming the Earth dangerously, although other experts weren't so sure. Delegates of 160 countries—most of the countries in the world—had agreed to negotiate a treaty requiring all countries to participate in a complicated scheme to reduce the world's total emissions of CO_2 from factories, power plants, motor vehicles, wood fires, and other sources.

From what the media reported, any casual consumer of the news would likely have drawn the following conclusions:

- Scientists are divided on the issue: some believe the large quantities of CO_2 released by human activity are causing global warming, but others aren't so sure;
- Even if warming is occurring, it's not necessarily a big problem, and if restricting CO_2 would hurt the economy, then that really isn't warranted at this time;
- *If* warming is a problem, the climate treaty will take care of it;
- This issue is no more urgent than a long list of others clamoring for our attention.

These inferences could hardly be more mistaken. In fact, while scientists are independent-minded people who normally argue with each other endlessly about details of methodology or interpretation of their findings, by the time of the Kyoto meeting they had formed an extraordinary worldwide consensus. A task force of leading climate scientists from 98 countries, known as the Intergovernmental Panel on Climate Change, or IPCC, had studied the problem exhaustively and had issued a 1995 report warning that this is a problem of enormous consequence. The report had been authored not by one or two leading researchers as most scientific studies are, but by 78 lead authors and 400 contributing authors from 26 countries, whose work was then reviewed by 500 additional scientists from 40 countries, and then *re*reviewed by a conference of 177 delegates representing every national academy of science on Earth. The IPCC had been unequivocal in its conclusions that (1) warming is happening, rapidly; (2) human activity is causing it; (3) the warming is likely to unleash devastating weather disturbances ranging from unnaturally heavy storms and floods to heat waves and droughts; and (4) it is therefore urgent that carbon emissions be cut sharply all over the world, but particularly in the industrial nations where these emissions are heaviest.

If all this got largely lost in the news surrounding Kyoto, there was one other conclusion that got completely buried: the agreement, as drafted, would actually do little to stop the world's climate from becoming dangerously disrupted. The treaty called for cutting CO_2 emissions from industrialized countries by a total of about 5 percent below 1990 levels worldwide by the year 2010 (7 percent in the heavily industrialized United States), but the IPCC scientists noted that to stabilize climate would require a cut of at least *60 to 80* percent.

It wasn't the climate scientists, however, who were to write the treaty. In a world of increasing specialization, that wasn't their job. The negotiations were done by politicians, bureaucrats, and diplomats. In the end, after years of wrangling (including another global conference in Buenos Aires a year after Kyoto), the negotiators failed to come up with any credible way of averting the threat this carbon spike poses. They failed because they were acting under the guidance *not* of climate scientists, but of an international coalition of oil, coal, electric power, automotive, and chemical industry interests whose products are the main sources of the rising levels of CO_2. They are the same industries that have dominated the global economy for the past century and that apparently intend to continue that domination far into the *next* century—even if doing so means waging a kind of war.

Despite such periodic spasms of publicity, most people have paid little attention to greenhouse gases, and it's not hard to see why. Carbon dioxide, after all, is a normal constituent of the atmosphere, and in that capacity it is quite harmless. In fact, it's essential to life. As most of us learned in elementary school (and were repeatedly reminded by oil and coal industry representatives during the climate treaty debate), the CO_2 exhaled by animals is as necessary to the growth of plants as the oxygen released by plants is necessary to the breathing of animals. Carbon dioxide and oxygen are inseparable partners for life.

When the concentration of CO_2 began to rise sharply about two centuries ago, no one was really aware of it. During the nineteenth century, a few scientists began speculating that increased carbon emissions (mainly from heavy burning of coal) could increase solar heat retention, but these speculations attracted little interest. In the mid-twentieth century, when the rise was accelerating sharply, our minds had been abruptly turned to a far more charismatic atmospheric worry: we'd be rained on by the fallout of nuclear bombs. Almost no one dreamed that the quiet rise in carbon dioxide could eventually induce a kind of violence that rivals even nuclear war.

Excess CO_2 prevents the Earth from radiating heat at its normal rate. It's like making someone run in a rubber suit on a hot day, which could quickly cause that person to overheat and develop disrupted patterns of breathing, sweating, and behavior—and eventually, spiraling core temperature and death. The Earth's response to overheating is a disruption of its normal patterns of wind and precipitation, and the power of that disruption can be staggering. "An average hurricane represents the energy equivalent of half a million [Nagasaki-sized] atom bombs," wrote climate researcher Lyall Watson in his 1984 book *Heaven's Breath: A Natural History of the Wind*.

That comparison may be theoretical, but the point is valid: our most awesome weapons still pale beside the potential powers of the living planet, and it will not serve us well to unleash those powers unwittingly. Fortunately, hurricanes typically expend most of their energy precipitating water over open ocean and rarely strike cities head-on. But that will change, at least by degrees, as cities spread out over more of the planet's surface, and as hurricanes get larger and more frequent. At the time Watson wrote his book, the largest natural disaster ever to hit the United States had been Hurricane Betsy in 1965, which produced about $1 billion worth of damage. In 1992, the damage inflicted by Hurricane Andrew amounted to $25 billion. Even allowing for inflation, Andrew was ten times as devastating as the previous record holder. Since Andrew, total annual storm damages worldwide have quadrupled. And most climate scientists agree that this may be only the beginning. Indeed, after Hurricane Mitch left much of Honduras, Nicaragua, and Belize in ruins in 1998, BBC News reported that the storm had "blown away the scale scientists use to measure severe storms."

That the air we exhale should have such power defies credulity. But it's easy to forget that in both plant and animal physiology, small tolerances often spell the difference between health and disease—or life and death. Vitamin A, white wine, and sleeping pills are all benign or beneficial at the right times and doses, but lethal in excess. A one-degree rise in body temperature may be within normal range, but a seven-degree rise, if not quickly reversed, is fatal.

What's true for individual people or plants is true for the planet as a whole, which has many of the same kinds of balances found in individual organisms. The biosphere has evolved, over eons, a set of fine tolerances in the composition of its air and the temperature ranges of its ecosystems. Those tolerances shift naturally over time, and the planet's ecosystems evolve slowly with those shifts. If the fundamental planetary chemistry undergoes a large change *suddenly*, however, there's no time for evolutionary adaptation. The effect will be like that of a heat stroke in a rubber-suited runner—or of a dangerously high fever in a child, with no known medicine available to reverse it.

For an indication of just how fine the tolerances of the living planet are, consider the density of CO_2's partner, oxygen. At present, oxygen constitutes a little over one-fifth of our air (the bulk of the air is nitrogen). If the concentration of oxygen were to increase by 20 percent, say physicists, all the vegetation on Earth would burst into flame and virtually all life would be destroyed—most of it within hours. (Oxygen alone wouldn't cause the fire; it would take a spark. But of course, there are billions of sparks in the air at any moment.) Carbon dioxide accounts for much less of the air than oxygen and doesn't pose that *kind* of risk, but in the past half-century the concentration of CO_2 has already increased by about *30* percent. Graph 1 shows how sudden that increase has been.

The concentration of CO_2 began a slow rise with the global expansion of human development, which over the past several millennia has reduced the world's forest cover by one-half. Trees store much of the planet's carbon, and when trees are burned to make space for farming or to produce heat for cooking, the carbon is released. In recent centuries, as forest storage capacity declined, more carbon remained in the air—and the concentration rose. But the big surge began with the advent of the Industrial Revolution and the mechanization of activities once performed mainly by people's bodies, or by the bodies of other animals such as horses or oxen. Transportation, farming, mining, manufac-

1

THE CARBON DIOXIDE SPIKE

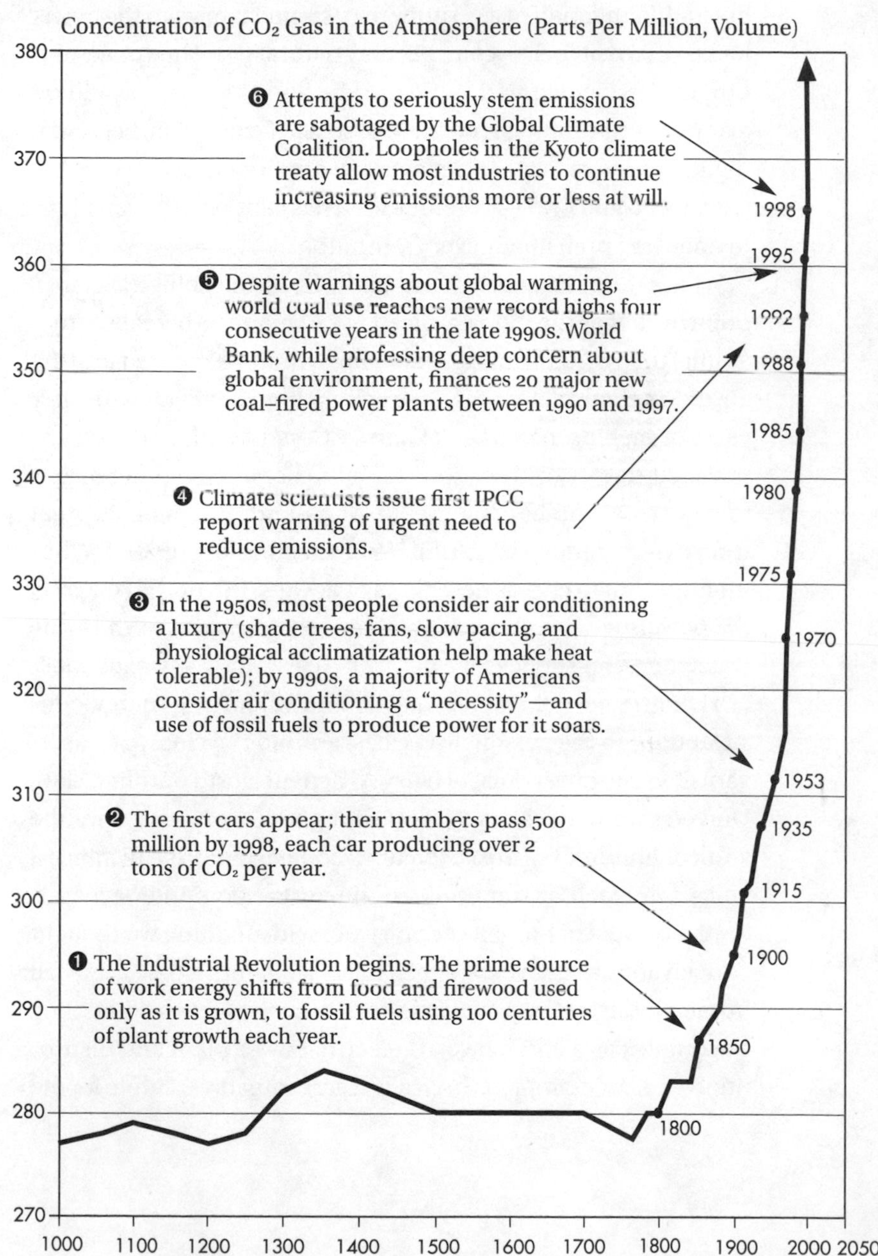

turing, and communications had all been done by animal power, but within the span of a century they shifted largely to the power locked in fossil fuels—coal, oil, and natural gas. The concentration of CO_2, which had increased by about 1 part per million every 400 years through most of our species' expansion, began rising 100 times as fast—by an average of 1 ppm every 4 years between 1800 and 1970—and since then has accelerated even more, to another 1 ppm about every 8 months.

What's interesting about this shift, from an ecological standpoint, is that as the industrialized world made its big move from animal to fossil fuel energy, the main byproduct emitted into the air didn't change. A woman working a hoe, or walking to carry yams or melons to market, exhales CO_2, and so does the diesel-powered tractor or truck that replaces her. So the difference is one of *scale*, and here's a great irony—and potentially a great tragedy—of industrialization. What technology does for humanity is not to give us new kinds of powers, but only to magnify the powers we already had. A car, for example, gives us the same transportation services we get from legs, but vastly expanded. "Driving 10,000 miles per year is equivalent [in energy consumption] to the personal services, around the clock, of 30 servants," wrote physiologist Henry A. Bent of North Carolina State University a few years ago. And so it goes: the tractor does the work of hundreds of arms; the telephone extends the distance a voice can reach by thousands of times; the computer expands on the powers of human memory beyond anything we thought conceivable a few years ago. These are simple cases, but all human industries are expansions of ourselves. Even something as complex as a coal-powered electric power plant and distribution grid, for example, is basically a gargantuan substitute for the legs, arms, and backs of peasant farmers gathering wood or dung for fuel.

These huge expansions of our natural powers have been cause for frequent celebration, as invention piles on invention. And as our powers have expanded so has our sense of ourselves as having *unlimited* powers. What's been overlooked is that the byproducts multiply just as impressively as the services do, and in fact multiply even more steeply, because the energy efficiency of oil- or coal-powered machines or electricity is far lower than that of the human-powered work we have increasingly abandoned. So, while the output of our industry is multiplied hundreds or thousands of times, the amount of waste it produces—including CO_2 and heat—is multiplied even more.

In many respects the benefits of the techno-multiplication may be unarguable, but in other cases it is completely irrational. For example, consider the man who mows his average-sized suburban lawn with a gasoline-powered riding mower, then drives a gasoline-powered car to the local gym, where he pays good money to walk on a Nordic Track machine and exercise the very muscles he'd have used to push a hand-powered mower.[1] Is it that this man is being taken for a ride in more ways than he realizes, or is it that his ability to integrate knowledge has been severely underdeveloped or damaged? It's a critical question, because this kind of misguidance now drives our lives in uncountable ways.

1 ❧ Tests by the US Environmental Protection Agency show that the average power mower emits, in one hour of operation, the same amount of hydrocarbons that a 1992 Ford Explorer emits in 23,600 miles of driving, reports the environmental labeling organization Green Seal.

If we increase the temperature of a human body by 20 percent, or the oxygen in our air by 20 percent, we know what happens. The result is quick death, whether to the individual or to the biosphere. But what happens when we encounter an even larger increase in carbon dioxide?

That was the question studied by the scientists of the IPCC. Their first report had come out in 1992, and it was that report—little noticed by the public—which alerted the world's fossil fuel industries that their domination of the global economy might be in jeopardy.

Soon after that report came out—several years before Kyoto—these industries launched a countercampaign of unprecedented sophistication. Centered in the United States but with tentacles worldwide, it played on two kinds of fears to which people were becoming increasingly vulnerable in the age of globalization: fears of economic setbacks and of cultural invasion. [This was also the period during which the North American Free Trade Agreement (NAFTA), the European Union (EU), and the Asia Pacific Economic Cooperation Forum (APEC) were coming to power, and many people were nervous about losing their jobs either to immigrants arriving or employers departing.] What made the campaign especially insidious, and hard to discredit, was that both of those threats are real. The world has entered economically precarious times, and cultures on every continent are being homogenized and diluted.

In the mid-1990s, more than a hundred major corporations in the coal, oil, utility, motor vehicle, and chemical industries, among others, assembled a formal lobbying organization, the Global Climate Coalition (GCC), to conduct this campaign. All told, the GCC probably wielded more power over the global economy than even the OPEC cartel that had established an economic stranglehold in the 1970s, but the GCC was far less visible —it was made up of transnational companies, not countries.[2] In the year preceding Kyoto, however, the GCC placed a $14 million barrage of ads in US media, warning that if governments took actions to reduce CO_2 emissions (which would mean in some way limiting sales of coal, oil, and gas), those actions would "cripple the US and global economies."

That assertion, for which no substantiation was offered, diverted attention from the question of what risk we would take by *not* reducing emissions. In the United States, TV commercials warned that requiring industry to reduce emissions would cause gasoline prices to rise by 40 cents a gallon. The ads did not mention the possibility that the *alternative* to that modest risk might be to impose a far greater risk—of devastating increases in flood or hurricane damages, shrinking water supplies, or the pervasive ecological disruptions that could be triggered by rising temperatures. The possibility of higher gas prices was just annoying enough to keep those larger risks out of the debate.

The fear of cultural subversion was subtler. The essence of it was that if a global treaty were imposed, the sovereignty of individual nations would be threatened. The GCC campaigners took particular care to arouse the fears of US politicians, some of whom were known to be deeply hostile to any "globalist" developments (such as the growing influence of the United Nations) that might raise challenges to US sovereignty. The Kyoto agreement called on the industrial countries, which were producing the heaviest CO_2

2 ❧ These companies are both everywhere and nowhere. They are at every gas station and electrical outlet, yet unlike Saudi Arabia or Venezuela, they do not appear on any map. They are more secretive and generally less accountable than governments—as chapter 7 will discuss.

emissions, to take the lead in cutting them back. In those countries, the output had reached more than three tons per person, per year—compared with only half a ton per person in the developing world. But just before the convention, the GCC began asking, in a final blitz of newspaper and TV ads, "why should developing countries, like China, be given *lesser* obligations?" The ads pointed out, correctly, that rapidly industrializing China, with its huge population and vast reserves of coal (the worst of the carbon-emitting fuels) could soon produce quantities of carbon gas that would obliterate any gains made by US or European cutbacks. It was a persuasive argument, with the issue of "fairness" successfully diverting policymakers' attention from the question of how to avert the consequences the scientists had warned of.

This effectively aroused anxieties in both the developed *and* developing countries—instigating rancorous arguments between them and further distracting them from the thing that threatened them both. The US politicians didn't want to get caught obligating their country to carry more than its share of the burden. The developing countries didn't like the thought that hostilities toward them were being aroused in rich countries, and that this might result in restrictions that would inhibit their future economic development.[3] One of the most bitter responses

3 ❧ "We cannot allow a world in which some countries have to freeze their carbon dioxide emissions at one level and other countries at another level," wrote Anji Sharma of India's Centre for Science and the Environment in a letter to Prime Minister I.K. Gujral three weeks before Kyoto. "This would mean freezing global inequity." And an editorial in *China Daily* stated, "There are those who are unwilling to see China progress and who are trying to contain its development by pointing fingers at the world's environmental problems."

to the US position came from Kinza Clodumar, president of the tiny Republic of Nauru, an island in the South Pacific that will be particularly vulnerable to any rise in sea level produced by global warming. (The rise is coming both from partial melting of polar ice caps and from thermal expansion of the ocean water.) The industrial countries have produced the bulk of the CO_2 so far, Kinza pointed out, and if they don't take the lead in reducing it, countries like his face "a terrifying, rising flood of biblical proportions." This would amount to "willful destruction of entire countries and cultures" and would constitute "an unspeakable crime against humanity," he told delegates to the Kyoto conference. This wasn't exactly what the GCC wanted to hear. But by provoking such rancorous rivalries among governments, its campaign effectively kept the world distracted from the real problem: that it didn't really matter who bore the brunt of a 5 percent cut if plans weren't soon made to phase out the world's coal- and oil-based energy industries *altogether.*

When the Kyoto treaty was signed, environmentalists cheered. The World Resources Institute called it "an historic step in the history of humanity." The Global Climate Coalition appeared to have been defeated, because it had loudly called for not signing, but the delegates had signed nonetheless. But in the weeks that followed, it became clear that the GCC had won. The treaty would still have to be ratified (Kyoto came up with the wording, but not the legally binding commitments), and the fears of economic disaster were now beginning to fester. Another convention was set for Buenos Aires, Brazil, in late 1998, to work out details of implementation. The negotiations leading up to this were voluminous enough, if fully transcribed, to fill a dump truck. But all those negotiations were over procedures that could never solve more than 5 percent of the problem and are very unlikely to end up doing even that.

Almost forgotten, in all the fuss over the climate treaty, was a far more meaningful and less compromised document that had been written five years before. In 1992, the same year the climate scientists of the IPCC issued their first report, a broader assemblage of 1,670 of the world's most accomplished scientists from *all* fields issued a joint statement, the *World Scientists' Warning to Humanity.*

The *Scientist's Warning* was concise, written in English as the most accessible language of international communication. At the time, much less was known about the mechanisms and magnitudes of climate change than is known now, but enough had been discovered to impel these scientists to take extraordinary action. In their introduction, they wrote:

> "Human beings and the natural world are on a collision course."

The statement listed the specific kinds of human activity that are courting danger and ended with a section entitled "What We Must Do," that listed five principal declarations. The first one began: "We must . . . move away from fossil fuels to more benign, inexhaustible energy sources to cut greenhouse gas emissions. . . ."

The statement was signed by 104 Nobel-Prize winners in the sciences—a majority of all those still living. But though they would finally convince a few automotive and oil industry leaders that their industries really have come to a critical juncture, neither the scientists of the *Warning* nor those of the IPCC would be able to influence general public perceptions—or policies—the way the GCC hard-liners would over the next few years. In the year following Kyoto, the construction of new coal-burning power plants would continue in China as though nothing had happened; hundreds of new oil wells would be drilled in the oceans; millions of new gas-guzzling sport utility vehicles, vans, pickup trucks, and luxury cars would continue to be sold, most of them with even lower fuel efficiency than a decade earlier; and the spike of carbon gas would continue to rise.

THE EXTINCTION SPIKE

The second of the four megaphenomena appears on a graph of human history as the steepest spike and ultimately the most dangerous. Yet it, like the silent rise in CO_2, is largely invisible to the majority of us: it is a sudden, sharp rise in the number of species in the world experiencing population crashes and going extinct. Many of the species are disappearing without our ever noticing; as we go about our everyday lives, they are dying off in distant forests, deep ocean waters, or in the soil under our feet. Yet their disappearance threatens to unravel the web of life that *sustains* our everyday lives.

A graph of this spike, showing the rising number of species extinguished each year, is so steep that to plot it the way most scientists would, we would have to use a logarithmic scale.[4] On the other hand, to show the magnitude of the biological crash now occurring in its real proportions requires a graph far too large to fit on this page. If the vertical axis is spread out just enough to show the substantial rise in the rate of extinction that took place during the agricultural revolution, when people began to deforest the world in earnest, then the spike of the past few years would require a vertical foldout of this page long enough to reach from your lap to your ceiling. Each inch of this graph would probably represent more kinds of birds, beetles, monkeys, flowers,

4 ❧ In a logarithmic (or "log") scale, instead of each vertical step representing an equal amount of increase (from 10 to 20 to 30, 40, etc.), each step upward is ten times the one before (10, 100, 1000, 10,000, etc.).

fish, and other living things than you have seen in your whole life. This spike, greatly compressed in order to fit on the page, is shown by graph 2.

To put this second spike in even larger perspective, if the horizontal axis were stretched back for millions of years instead of the seven thousand shown here, the shape would be about the same—the line would show as fairly flat, except for a few low wobbles, all the way back to the time the dinosaurs were suddenly killed off 65 million years ago.[5] Using the same scale used here, the line would run across the bottom of a foldout graph nearly twenty miles long. Those earlier disruptions took place slowly enough to allow for many animal and plant populations to migrate or adapt, so many species may have survived then that will not survive now. Only now, as of the period since World War II—the last *inch* of that twenty-mile-long graph—does the spike abruptly shoot upward as it did at the time of the dinosaurs' extinction. What happened back then, most scientists believe, was not a natural climatic or geological change but a traumatic once-in-an-eon event—a collision between the Earth and a large asteroid. What's happening now, though few of us yet see the signs, is also a once-in-an-eon event.

Scientists have been studying this phenomenon of decimated animal and plant life, or biodiversity loss, with as much intensity and concern as they've been studying CO_2 and climate change. Not surprisingly, the two spikes turn out to be linked, in some important ways. The slash-and-burn clearing of rainforest to make space for oil-palm plantations or crops, as is now taking

5 ❧ Those wobbles would reflect earlier mass extinctions caused by natural climatic changes that occurred over hundreds of years, so the numbers of extinctions per year were much lower than they are now.

2

THE EXTINCTION SPIKE

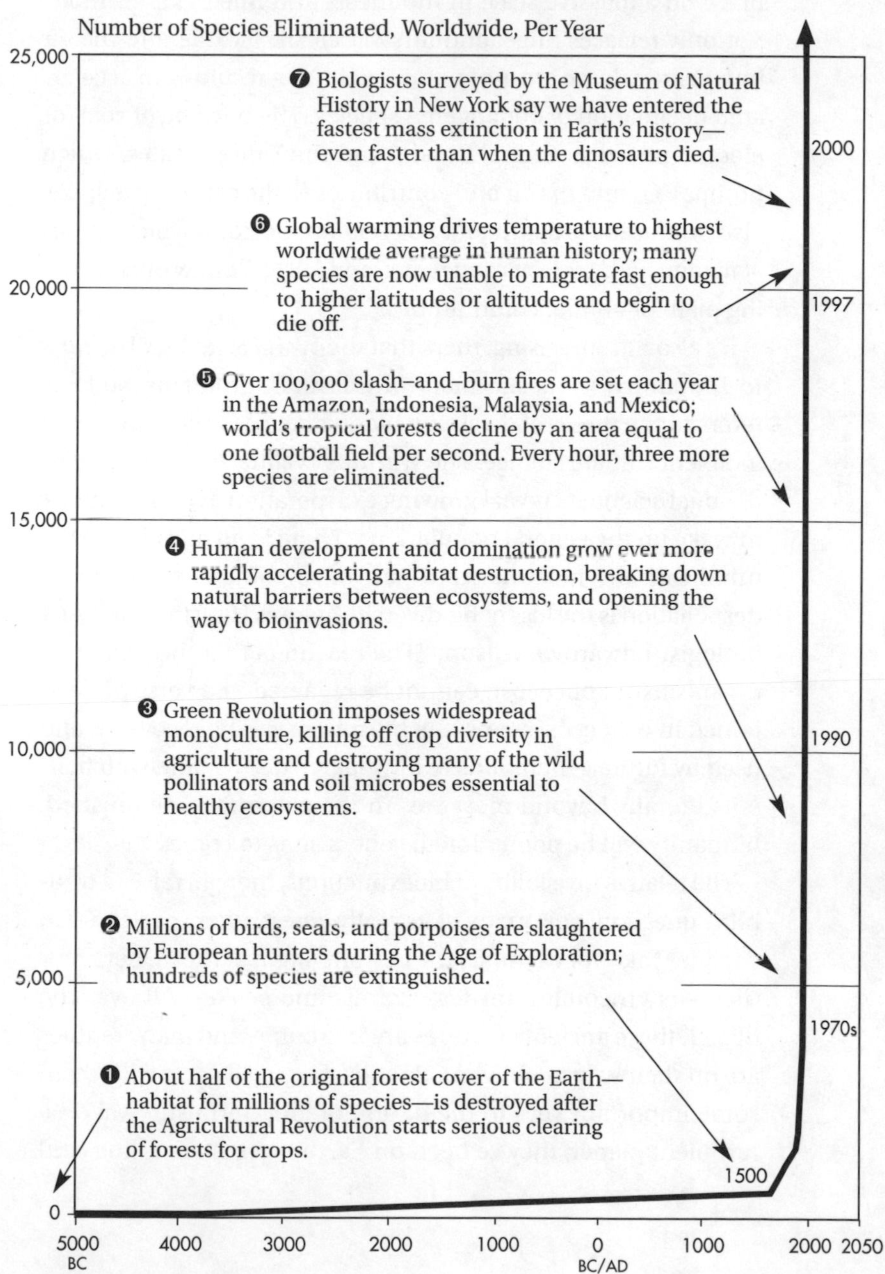

place on a massive scale in Indonesia and Brazil, for example, not only releases huge amounts of carbon dioxide into the air but destroys large areas of natural habitat—resulting in accelerated decimation of endangered species. The burning of coal for electric power in China, Russia, and the United States, which pumps CO_2 into the air and contributes to the carbon-gas spike, also produces the byproduct SO_2—sulfur dioxide—which causes acid rain, pollutes lakes and rivers, and kills off many once-thriving plant or animal communities.

It's also not surprising, then, that the *World Scientists' Warning to Humanity,* in its summary of the main problems we face, refers to the threat of declining biodiversity as well as to that of incipient climate change. The *Warning* is clinical in tone, but individual scientists reveal growing exasperation at their inability to wake up the general population. "There is no question in my mind that the most harmful part of ongoing environmental despoliation is the loss of biodiversity," writes Harvard University biologist Edward O. Wilson. "The reason is that the variety of organisms . . . once lost, cannot be regained. If diversity is sustained in wild ecosystems, the biosphere can be recovered and used by future generations to any degree desired and with benefits literally beyond measure. To the extent it is diminished, humanity will be poorer for all generations to come."

The relative invisibility of bioextinctions, though, raises a troubling question: how many of us really *care* if other species are in decline? Take, for example, the current plight of the planet's reptiles—its crocodiles, turtles, lizards, and snakes. All over the planet, these ancient creatures are in trouble, and many reptiles are on their way to oblivion. Reptiles may well carry with them some important keys to the nature of long-term survival on a turbulent planet; they've been on Earth more than a thousand times as long as Homo sapiens has. Yet many people unhesitatingly kill any snakes they encounter, and many evidently take the view that if we're doing just fine without dinosaurs, we can do without today's reptiles as well. Does it really *matter* that extinctions are on the rise?

Maybe it will seem to matter more to us if we note that it's not just reptiles that are in jeopardy. Amphibians, fish, insects, and birds, as well, are dying. Of the 9,600 species of birds in the world, 70 percent are in decline. According to Birdlife International, the UK-based organization that compiles bird census data worldwide, at least 1,000 species of birds are headed for extinction. Coming closer to home, a large proportion of all *mammals* are threatened. Many of us have been galvanized by the bond we feel with such fellow mammals as tigers, elephants, wolves, and whales, most of which are now in trouble. But then, narrow the focus even more and consider that order of mammals we call primates, to which we are genetically closest. One species of primate—humankind—has exploded in population, while the 232 others, taken as a whole, have plunged.

Even that, however, may not be alarming to those who are not moved by the faces of baby chimpanzees. To many, the ascendance of man is an evolutionary triumph: we have outcompeted the other primates and the other mammals. We have, as the Bible commands, subdued the Earth.[6] Competition has become an ideal for both evolution and economics. There are, no doubt, some businessmen who would far rather make another corporate acquisition than save the gorillas from extinction, and there are

6 ❧ The Bible also commands, "Hurt not the earth, neither the sea, nor the trees." (Revelation 7:3)

basketball players who would rather sink a title-winning shot than save all the whales in the oceans from final oblivion.[7] And these may be good people. They just don't see the point of winning the world only to have to start giving it back.

There are really *two* points that have tended to elude such born "winners." The first is that the great variety of species on the planet is our natural genetic capital, on which evolution itself depends. Evolution is the process that got us to where we are, and it is what will take us to wherever we will go in the future. But evolution depends on the availability of a vast diversity, both of species to form healthy ecosystems and of gene pools within species to help them adapt to prevailing conditions. The competition we so admire brought us to our present dominance because we were able to draw what we needed from those resources, whether in ourselves or in our surroundings. To severely diminish diversity in either of those realms (to replace diverse natural forests with monoculture plantations, for example, or to allow our own nature to be pushed toward greater uniformity through aggressive marketing of genes to parents the way Monsanto markets bioengineered seeds to farmers) could dangerously limit our future capabilities to stay healthy and to adapt. And those are capabilities we will need more than ever before as we undergo the climatic and biological disruptions that lie ahead. Those who are caught up in the pursuit of short-term competitive success may not care about this. And the number not caring seems to be growing, as attention spans grow shorter, the thrall of commercial profits grows more pervasive, and our abil-

7 ❧ "The dinosaurs went extinct, and I don't miss them," said Ruben Ayala, a California state senator, in explaining why he opposes legislation to protect endangered species for which he sees no use.

ity to connect with what happens beyond our immediate lives diminishes.

But there's a second point that's likely to hit home with even the most self-centered. The variety of species is important not only to long-term evolution but to short-term stability of ecosystems, *including those which are the sources of human food.* Variety is what enables agriculture to develop natural defenses against pests and diseases. The great potato famine of nineteenth-century Ireland might have been averted if varieties carrying genes resistant to that blight had been available—and when the blight was arrested, it was partly because blight-resistant varieties were found in time. Since then, however, the blight has returned to potatoes just as malaria and tuberculosis have returned to humans. Outside of Ireland, hundreds of different potatoes once provided genetic adaptability to all sorts of climates, soils, seasonal conditions, and diseases. But when McDonalds and other fast-food chains began their global spread they demanded, in each country they entered, that the varieties of potatoes cultivated by local farmers be abandoned and replaced with their global standard, the Idaho russet, so that a uniform procedure for processing French fries could be used everywhere. Today, the global potato harvest is precariously homogenous—heavily dependent on pesticides for protection. Yet pests are becoming increasingly resistant to pesticides. The fungicide metalaxyl was used to kill the potato blight for years, but the new strains of the pathogen have become more aggressive and resistant to the fungicide. This time, the blight has attacked potato crops not just in one country but throughout North and South America, Europe, Africa, and Asia. Worldwide, production has fallen by 15 percent. Scientists in Peru are trying to develop a more blight-resistant potato, but success in that venture would only lead to still more extensive monoculture, which in the battle against pests is ultimately a losing game.

This is happening not just with potatoes, but with rice, wheat, and a wide range of other foods. At first, it results from a deliberate strategy of genetic conformity, as with McDonalds. But that then sets up a vicious cycle, for as more indigenous varieties disappear, more pressure is put on those few that remain. According to records kept by the US Department of Agriculture, for example, at the beginning of the twentieth century there were 307 varieties of sweet corn being grown in the United States, but by the last years of the century fewer than 40 of these remained.[8] Garden beans dropped from 578 natural varieties to 32; spinach from 109 to 7. "Loss of genetic diversity in agriculture is leading us to a rendezvous with extinction—to the doorstep of hunger on a scale we refuse to imagine," write biodiversity activists Carey Fowler and Pat Mooney.

Since 1900, reports the Rural Advancement Foundation International, about 75 percent of the genetic diversity in agricultural crops has been lost. And what is happening in the fields is also happening in our backup food source, the oceans. Overemphasis on high-volume production has decimated populations of once plentiful fish, so that fish once regarded as "trash" are now eagerly sought for human consumption. The populations of oceanic fish are no longer keeping up with human demand. As these populations fall, their variety too is declining.

The evisceration of biological wealth is happening across a

8 ❧ Some new engineered varieties were introduced in recent years, but these are dependent on heavy applications of chemical fertilizers and pesticides to be effective. They may temporarily increase gross productivity, but do not protect the natural diversity and stability of the farm environment.

range of ecosystems—in rivers and lakes as well as well as oceans; in forests, where complex ecosystems are being razed and replanted with homogenous tree plantations; and even in the soil itself, where many of the microbes essential to healthy plant growth are being killed off by agricultural chemicals and pollution. Until very recently, little was understood about just how important natural systems are to human economies—and even to human survival. But in 1997, a study published in the British journal *Nature* found that the economic value of ecological services—such as soil formation, water filtration, biological control of pests, and production of oxygen—far exceeds the value of the entire human economy. Bees, and other wild insects, for example, pollinate about 80 percent of the world's major crops, other than grains—and without them we'd lose many of our foods. But now, with their meadow habitats rapidly being paved over or bulldozed for spreading human development, the bees, too, are disappearing.

All life on Earth depends on solar energy for survival, either directly (through photosynthetic growth of plants) or indirectly (through consumption of plants or of other animals that eat plants). In 1998, the International Union for Conservation of Nature issued an 862-page report, the *IUCN Red List of Threatened Plants*, summarizing 20 years of research by 16 scientific organizations. The study found that about 34,000 of the *known* species of plants on Earth are now approaching extinction—and only a small fraction of the planet's species have been catalogued.[9]

Around the same time, the American Museum of Natural

9 ❧ While plants constitute only a small minority of the millions of species on Earth, they are critical because without them no animal life can exist.

History in New York conducted a survey of 400 experts in the biological sciences. They included researchers in biochemistry, botany, conservation biology, entomology, genetics, marine biology, and neuroscience. The survey found that a large majority of the scientists believe that during the next thirty years, one of every five species alive today will become extinct. A third of the scientists predicted that as many as half of all species will die out in that time. The consensus was that the Earth is now in the throes of the fastest mass extinction in the planet's history—which would make it even faster than when the dinosaurs died.[10] At the same time, the museum conducted a parallel survey of the general public, with a curious finding: most people were unaware that we are in the midst of a biological crash—and that it is a crash we have brought upon ourselves.

THE CONSUMPTION SPIKE

Of the four megaphenomena that will define our time, the rise of unsustainable consumption is the most visible yet the most difficult to measure. Unsustainable consumption is not synonymous with "conspicuous" consumption, though it includes it. But more problematically, unsustainable consumption includes forms of activity that to most of us seem perfectly normal. And what is "normal" to us will probably be regarded by humans of future eras as pathologically excessive. We are practicing a kind of commerce that is drawing down the Earth's finite resources—its topsoil, water tables, and genetic resources—far faster than natural processes can regenerate them. Glancing back at the ex-

10 ❧ The full ecological effects of the asteroid collision may have taken centuries to play out, whereas the main impacts of the current rate of loss may be seen within our own lifetimes.

tinction spike, for example, note that Harvard's E.O. Wilson has estimated that we are consuming genetic resources between 1,000 and 10,000 times as fast as evolution produces them. Trying to see a clear, measurable trend in this is like watching a manic-depressive on a spending binge and trying to predict—without knowing much of anything about his bank balance, assets, or debts—just how long he can keep this up before going broke.

Yet try we'd better, because a number of signs say our spree can't last. Among them:

- We consume trees for lumber to build houses for the expanding human population and to make paper and packaging for our expanding volume of purchases. And yes, to make the paper for this book. Even when we're not using the wood, we cut away the world's tree cover as though it were weeds. Australia, for example, is largely denuded; yet in the province of Queensland alone, what little forest remains is being razed at a rate of 60 soccer fields per hour. But trees can't grow as fast as our demand is growing, and the world's forests as a whole are diminishing at a rate that the Worldwide Fund for Nature now calculates to be 1 percent a year.[11] Moreover, an amount of cutting that reduces forested area by 1 percent this year will reduce it by a slightly larger percentage next year, as

11 ❧ This does take into account the much-advertised replanting of trees by companies like International Paper and Champion International. When a tree plantation displaces a natural forest, the forest ecosystem is destroyed and most of its species are killed off. But even if the plantation is counted as "forest cover," the planet's cover is declining by 1 percent a year.

the principal diminishes. And, as the economies of fast-industrializing countries like China and India boom, the demand for timber is exploding. At the projected rate, which includes both the consumption of wood for timber or fuel and the clearing of forest for farming, the world will be denuded of natural forest within the lifetimes of those who are now in their 20s or younger. A denuded planet cannot support human life. Of course, we won't *let* the planet be denuded; more likely, we'll replace natural ecosystems with highly simplified artificial ones—such as eucalyptus tree plantations and catfish farms. But doing that drives up the spike of extinctions still faster—and subjects all of humankind to a high-risk biological experiment. As with any experiment, it's a procedure for which *we can't know the results in advance*, and if it turns out badly we will not have a second chance.

- We consume topsoil—the stuff from which all French fries, tacos, and steaks are produced—much faster than the processes of nature can remake it. The consumption takes at least four basic forms: depletion of soil nutrients, contamination by salt or toxic chemicals (in effect using up soil as a dump for waste), covering over with pavement or buildings, and erosion into saltwater estuaries or the oceans, where it is no longer usable for agriculture. Each year, about 30 billion tons of the world's topsoil is lost to erosion. (The loss from Iowa alone would fill 165,000 Mississippi River barges a year, reported the environmental journal *Buzzworm* a few years ago.) Irrigation eventually builds up toxic levels of salt in billions more. Losing topsoil has about the same effect on a terrestrial community as losing blood has on person. Only so much can be lost.

- We consume fossil fuels at a rate that will—via the carbon gas spike—cause the ice caps to break up, the seas to rise, storms to worsen, pests to spread, and many ecosystems to die if they can't migrate fast enough to higher latitudes or altitudes. Forests that have evolved in specific temperature ranges, for example, can gradually migrate as winds and animals disperse their seeds—but they can't reproduce fast enough to keep up with the accelerated rate of climate change now being brought by global warming. The problem is not, as once thought, that we will run out of oil, but that we will run out of time.

There are more of these signs, but lest we simply become numbed or embittered by them, consider that our level of consumption is not actually a very good measure of the quality of life. In the past half-century, we've come to think of growing consumption as being synonymous with a growing economy and a rising standard of living. But if we think of a healthy economy in terms of the *services* it provides, rather than the *stuff* it provides, a very different picture emerges: we discover that it is possible to continue raising the quality of life—and even of the services provided by technology—without continuing to increase consumption. A few examples illustrate the point:

- A computer today easily provides hundreds of times as much capacity as one of equal size did a few years ago. So, the amount of *material* used to provide the same *service* has been greatly reduced. Or, put the other way around, we did not have to consume ever-larger quantities of silicon, plastic, and steel to produce ever-larger output in what computers can do. What we did with computer power, we can now do with the generation of electric power.

- A laminated I-beam for new house construction can be made with one-third as much wood as an old-fashioned 2-by-10 joist, yet be just as strong. Hundreds of builders are now using them. Instead of producing more living space by cutting more trees, it's possible to achieve the same end by using more intelligent design.
- An urban rail system uses only one-sixth as much energy per passenger mile as a commuter's car and generates only one-eighth as much CO_2 per ton of goods moved as a heavy truck. It uses only a fraction as much land.
- An electric lawn mower, by replacing a gasoline-powered one, eliminates 99.9 percent of the hydrocarbon exhaust—while saving 73 percent in total energy costs.

The list could go on to hundreds of items, most of them never mentioned by civic leaders or known to the public. But the point is proven strategies are available for reducing unsustainable consumption.

Meanwhile, however, the binge continues—and worsens. In the global economy as it is now, rising expenditures usually mean rising levels of unsustainable consumption. And, while measuring such consumption is hard, a reasonably good indicator of it is the index called Gross World Product. The amount of "product" produced by all the nations of the world, added up, is the GWP, and in the past few centuries it has soared, as shown in graph 3. David Korten, in his 1995 book *When Corporations Rule the World*, wrote that "we have added more to total global output in each of the past four decades than was added from the moment the first cave dweller carved out a stone axe up to the

3

THE CONSUMPTION SPIKE

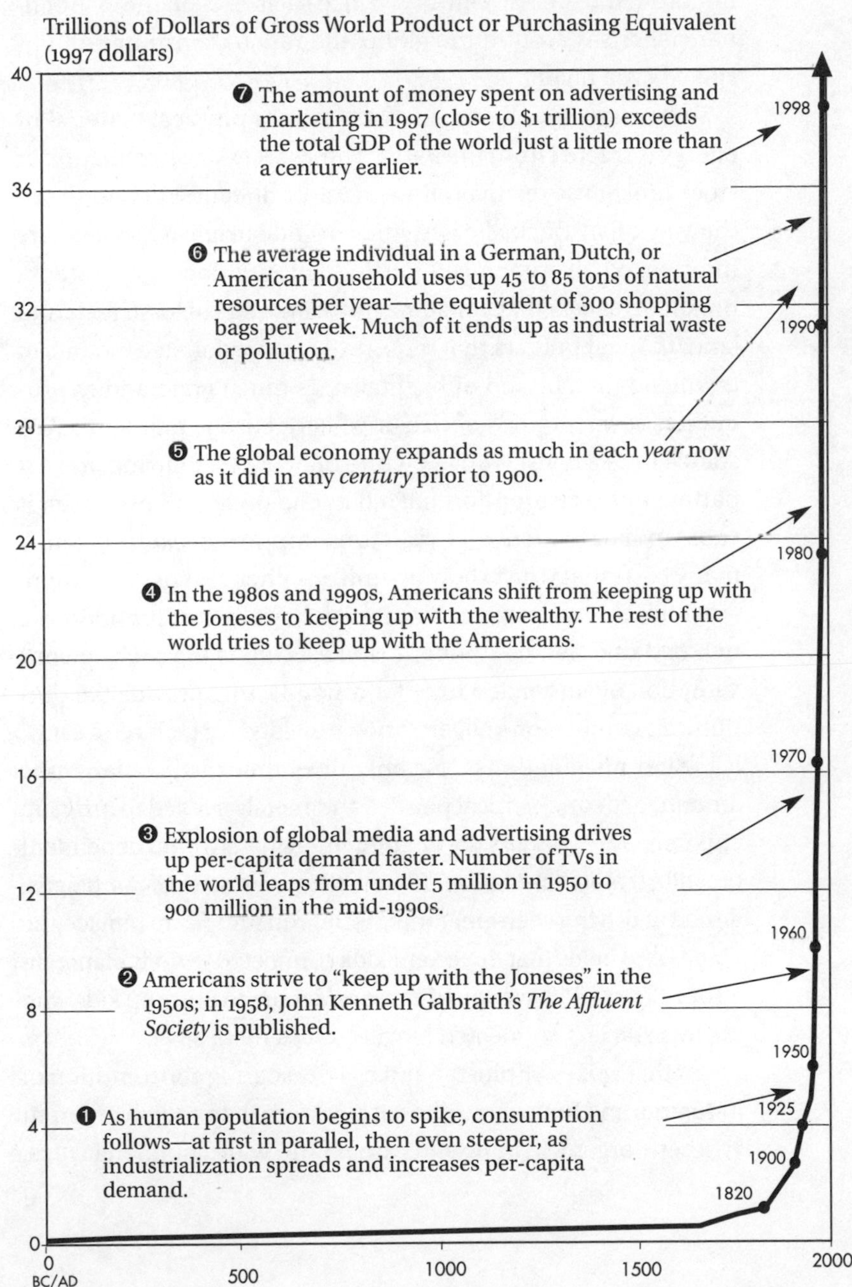

middle of the present century." Since then, even in the face of financial crises around the globe, the rate of consumption has spiked even higher.

To be sure, not all "product" consumes physical material or energy extracted from the Earth. The services of a psychiatrist or stock broker use relatively little material directly. But because of the way all of the basic activities of industrialized society are interlocked, everyone is a participant: the doctor and stockbroker drive cars that consume huge amounts of fossil fuels and land; they eat burgers that require disproportionate amounts of grainland (one pound of beef takes as much grainland to produce as seven pounds of rice or wheat). For the moment, GWP stands in as a conservative picture of how destabilizing present patterns of consumption have become. In actual numbers, it substantially *understates* real consumption because it omits many costs that won't show up until another year or generation.

Within this megaspike of materials/energy consumption is a nail bed of other, component, spikes. One is the soaring consumption of automated toys, for example, that provide the propulsion, conflict, or imagery once provided by children's arms, legs, and imaginations. Not only does that vastly enlarge the amount of petrochemical plastics and metals needed to bring up children, but it renders the children more passive and dependent on still greater stimulation. Billion-dollar marketing campaigns, aimed at driving ever-greater material consumption, replace the woods and fields that once kept kids connected to their planet. In a Toys-R-Us world, we spend more and more to bring up kids who are less and less connected to what keeps them alive.

Another spike-within-the spike is the soaring consumption of industrial and household chemicals. More than 70,000 different synthetic organic compounds are now in wide commercial use, most of them now at large in our air, water, and soil. Many were developed unnecessarily; they were poured into the escalating war on pests, or into a market for cosmetics that is built around the message that women can't be attractive or successful *without* cosmetics. Meanwhile, the spread of organic chemicals has been increasingly associated with a wide range of human health threats, including reproductive diseases, deformities in newborn babies, reduced IQs in children, and reduced sperm counts among men, as well as various cancers.[12] The vast majority of these chemicals have never been tested for their long-term effects on either human health or the environment. Of course, that was a key warning of Rachel Carson's landmark book *Silent Spring*. In 1962, when that book came out, the US chemical output had rocketed from 0.1 billion kilograms per year in the 1930s to about 40 billion kilograms. Today, it is in the vicinity of 200 billion kilograms and rising.

THE POPULATION SPIKE

With each passing month, the four megaphenomena become more entangled by feedback loops through which they all exacerbate each other. But the spike of human numbers, no longer controlled as it once was by diseases—and no longer limited in its impacts to the capacities of our own bodies—is the primary driver of the others.

12 ❧ A graph showing the consumption of these chemicals in the United States alone, if 1 inch of rise represents 100 million kilograms of chemicals consumed in a year, looks as follows: If we start at 1920, the line begins ¾ of an inch from the bottom axis and stays more or less at that level for the next 15 years. Then between 1935 and 1940 it rises to 100 inches, and between 1945 and 1995 it rockets to the height of a 9-story building.

For many people, hearing the topic of population raised as an urgent issue may seem almost like entering a time warp. Wasn't that an urgent concern way back in the 1970s? And hasn't it faded? Indeed, there was an alarm raised following the publication of Paul and Anne Ehrlich's *The Population Bomb* in 1968. At the time, most African countries had rampant fertility rates—some countries averaging 8 or 9 children per woman. But in subsequent years, a number of stabilizing measures have been taken. China has used draconian methods to curb its growth (though there are now some indications that it is creeping back up), and Africa has reduced birth rates. The United Nations Development Programme (UNDP) finds that in the developing world, growth rates have dropped. So, what's the problem?

To begin with, there can be danger in assuming that falling growth rates are always good news. Population growth can decline because people have become so deeply demoralized that they do not want to bring children into the world—as has happened in Russia. Or it can decline because people are dying of AIDS, as has happened massively in Zimbabwe and southern African countries. Sooner or later, if population growth isn't deliberately arrested by declines in births, it will be catastrophically arrested by rises in deaths—whether by disease, starvation, or violence.

Beyond that, the problem is that until the global growth rate drops to *zero,* the number of people continues to expand and the growth continues to compound. And, while rates are lower than they were a few years ago, they're still high enough to drive up the total number of humans on the planet by 80 million each year—the equivalent of having to find jobs, food production, fresh water, schools, plumbing, trash collection, hospitals, police protection, and sheer *room*, for seven more Beijings or Calcuttas this year, then seven more next year—70 more such megacities in the next decade.

To put that in perspective, consider the history of human population growth before our time. Shortly after Homo sapiens began to emerge as a distinctive species around a thousand centuries ago, there were an estimated 5 million individuals. Between then and the beginning of the Age of Exploration in the sixteenth century, the population grew to an estimated 500 million, or an average of roughly one-half million per century. Today, our population takes less than *three days* to increase by as much as it did in a whole century throughout most of our species' existence. The extreme suddenness of this increase, if we look at the "big picture" of our presence on this planet, is shown in graph 4.

This still doesn't seem to strike many people (outside certain academic and activist circles) as a problem, at least in the wealthier countries. One indication of that lack of concern lies in the way civil planners and economists still talk about "growth" as an unmitigated good. Many communities are fixated on "attracting business," which means growing in both population and consumption. And while a few places have seen grassroots no-growth movements emerge, such movements are still rare in the world as a whole.[13] For many planners, the time horizon is now so short that the construction jobs created by the building of new roads or housing developments are counted as part of the "growth" the community will experience—never mind that those jobs will disappear once the construction is complete, leaving

13 ❧ They also have little overall effect, because expanding population pressure breaks new ground one way or another, as inexorably as tree roots crack sidewalks and foundations.

4

THE POPULATION SPIKE

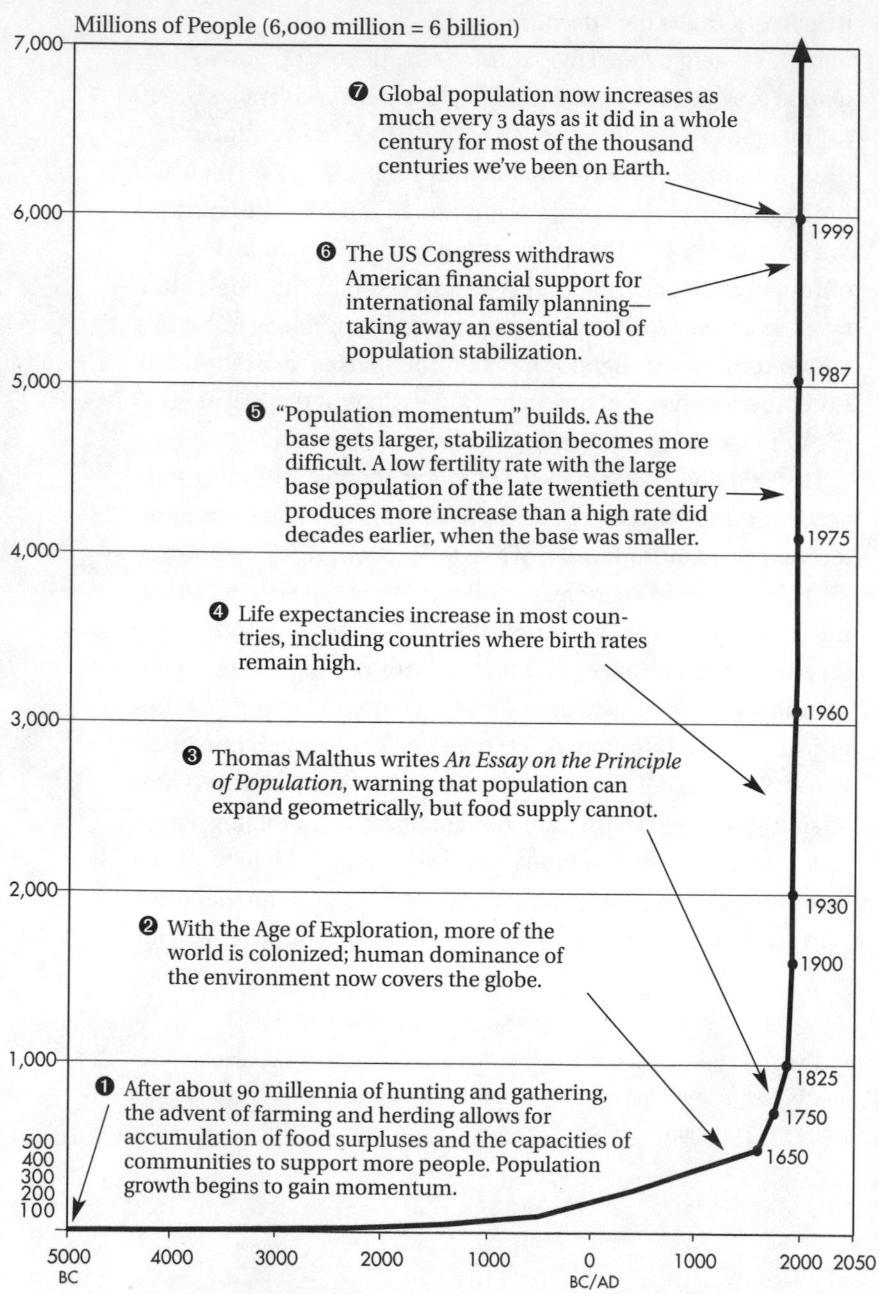

the community planners believing they have no option but to attract still more construction to replace the jobs that disappeared. It's a Ponzi scheme under the guise of public planning, and the net result is to encourage population to grow without accounting for how it can be supported in the long term.

To some extent, the lack of public concern is like the perceptual limitations of ants on a flood-borne log. In the developed world, the inundation hasn't yet pulled us under—or is only beginning to. And in already crowded places like Bangladesh or eastern China (where the inundation *has* begun), people are too focused on their immediate needs to know or care much about the world at large. The borders of nations that for the past several centuries have separated rich from poor, or educated from illiterate, have preserved an illusion of stability for a few last moments. But population growth brings some predictable changes, and the borders are now down in many places—and increasingly porous in all the rest. Human refugee and migrant flows are at their highest levels in history, and the tide will continue to rise. A decade ago, experts on population were warning that the planet's human carrying capacity was fast approaching its limits. Most experts in this field now agree that it has already passed those limits and we are living on borrowed or stolen assets we can never repay.

2

THE OBSTRUCTIONS TO PERCEPTION

In the first week of 1997, the Red River of the US northern plains, which forms the border between North Dakota and Minnesota, began to swell. At Grand Forks, North Dakota, it took a week to reach flood stage—28 feet above the normal water level. It continued rising, and by the third week fires were raging in the upper stories of buildings whose electric wiring or gas lines had been ripped loose, and whose lower stories were inaccessibly submerged. A World War II veteran remarked that the downtown district reminded him of Dresden, Germany, after it had been pounded by Allied bombers in 1945. But this was not World War II, it was the American midwest at the threshold of the twenty-first century. By April 28, the water had risen to 56 feet—the height of a six-story building—above its normal level, and the hulks of burned-out high-rise buildings stood eerily smoldering in the river. About 50,000 residents had been evacuated, their hometown trashed.

Described by flood experts as a "once-in-500-years event," the rising of the Red River followed hard on the heels of a number of other extraordinary weather disasters that struck in the 1990s—

each one covered briefly by the news media then swept quickly from public view:

1992 Hurricane Andrew hit Florida and wreaked greater damage than any storm to hit the region in 300 years. Experts said that if it had struck a little farther up the coast, it would have wiped out Miami like a nuclear attack.

1993 The Mississippi River brought its worst flood in 100 years, bursting levees the Army Corps of Engineers believed had been built for the ages. And Southern California was ravaged by its worst-ever brushfires.

1994 Eastern Australia suffered its worst drought in at least 100 years. So did Brazil.

1995 Spain reeled into the fifth year of its worst drought in history.

1996 China was struck by a flood that killed more than 3,000 people and left $26 billion in damages, breaking the record set by Hurricane Andrew.

1997 Forest fires destroyed vast tracts of tropical forests in Indonesia, Malaysia, and Brazil. The World Wildlife Fund reported that more of the planet's surface had burned in this year than ever before in recorded history. The most severe El Niño ever recorded began its Sherman's march across the planet.

1998 China was struck by an even more devastating flood than the one in 1996—this one leaving $36 billion in damages—exceeding the *entire world's* damages from natural disasters in any year prior to 1995. In the Yangtze River watershed alone, 56 million people were flooded out of their homes. During the same summer, Bangladesh

> was struck by a flood that left two-thirds of the country under water and 21 million homeless. Global average temperatures reached the highest levels ever recorded, and 45 countries were hit by crop-withering droughts. Of these, 21 were hit by both droughts *and* floods. The fires in tropical forests worsened and were seen in new areas: a large area of Mexico burned, sending smoke far enough to be visible in Chicago. In October, Hurricane Mitch killed an estimated 18,000 people in Central America. It was the most devastating year of climate-related catastrophes in recorded history.

There were many more of these disasters. Almost certainly, you were brushed by one of them yourself or know people who were. Invariably, though, they were treated as sensational, transitory events and then disappeared from public view. Some of the worst occurred during a period when US media were preoccupied with details of the White House sex scandal. And, there were hundreds of smaller weather anomalies that got almost no news coverage at all, including spates of tornadoes striking areas that had not experienced tornadoes within memory. In the first half of 1998 alone, US Weather Service data show that US communities were struck by an unprecedented 1,008 tornadoes.

It's dangerous to draw generalizations from individual incidents, of course. But records of aggregate weather-related damages around the world confirm that, in fact, a planetwide escalation—of the kind the world's climate scientists have warned about—*is* underway. One way to measure that escalation is to add up the financial damages caused by weather disasters. Christopher Flavin, a senior researcher at the Worldwatch

Institute, examined historic records of such damages worldwide. He found that in 1980 the total economic losses amounted to less than $2 billion, but by 1996 they had $60 billion for the year—a 30-fold rise.[1] By the time the Red River hit the upper Midwest in 1997, weather disasters had become so frequent that to many people they were coming to seem almost normal.

In the United States, that drift of perception is comparable to the drift that accompanied the growing frequency of fatal automobile accidents as people became increasingly car-dependent throughout the twentieth century. In the last decade of the century, as many Americans were dying in car wrecks *each year* as were killed, decades earlier, in seven years of Vietnam War combat. Yet, while the war carnage traumatized the nation, we are barely aware of the even larger magnitude of carnage on the roads. It is *because* those accidents are so commonplace that they rarely galvanize public attention. Car accidents are events people are resigned to; their frequency is not a public issue. In the rising frequency of floods, fires, and storms, a similar resignation is evident: while the drama of the destruction attracts momentary attention, the *pattern* of intensifying storm activity has come to be expected. Even among the victims, there are signs of widespread fatalism—a feeling that such disruptions are terribly unlucky, but unavoidable.

The first news from Grand Forks was very similar to that which has followed virtually every major disaster of recent years: a focus on the most graphic details of the event itself. Over several

1 ❧ The escalation would continue, and in 1998 the world's total storm damages would reach close to $100 billion as compared with an average of around $2 billion per year before the 1990s.

days, reporters followed these details closely: the height of the water, projected time of peaking, descriptions of property destroyed, number of people drowned, number of people evacuated, number who *refused* to leave, and some human interest interviews with rescue workers and evacuees.

After a few days, as is usual, there were follow-up stories going into greater depth, and these offered an opportunity for investigating the larger significance of the event. One such story in the *Washington Post*, about ten days into the flood, assessed the damage, the comments of local officials, and the plight of the residents. But there was something missing—something that had also been missing in similar "in-depth" assessments of most of the other 500-year and 100-year storms of recent years: there was no mention of *the probable role of human-caused climate change* in this flood. Of the 1,300 words of the *Post* follow-up, not one word touched on this probability, or on any of a number of critical issues that probability raises. And, though those issues are of far greater news importance than the flood itself, they were not pursued by such mainstream publications as *Time* or *USA Today* or the *New York Times* either. The unmentioned issues were these:

The cause: As the consensus of climate scientists makes explicitly clear, events of this intensity are not just acts of nature, but have been exacerbated by human activity—by the spike of CO_2 and other greenhouse gases. Human-caused warming triggers changes in the normal movements of wind and water that result in stronger storms, heavier rains, more precipitous melt-off of winter snow. (Increased warming causes more rapid evaporation from rivers and cropland, for example—loading more water into storm clouds and more snow into the mountains upstream.) Yet, none of the Red River victims were reported to be *angry*, at least not in the way people are often angry when they find that a pulp mill has been dumping dioxins into their drinking water, or that the state is planning to seize their land to build a highway. It was as though the misfortune had been just a random "act of God"—not for them to question. There wasn't the faintest indication, in the reported interviews, that those whose homes had been swept away might have any hard questions to ask about whether the rising severity of such destruction could have been mitigated years ago, if *people* had had the courage to act on what they knew. Or, more important, whether serious action is being taken now to avert such devastation in the future. There was no suggestion that any of the residents might now want to reflect on their *own* small roles in the increasingly aggressive human occupation of an increasingly irritable planet.

The chances of recurrence: Could it happen again in a few years? I would expect this question to be of surpassing interest, not only to those in Grand Forks and hundreds of other flood-vulnerable areas, but to taxpayers around the rest of the country who are footing the bills for mounting claims of disaster relief. For, in fact, the so-called 500-year or 100-year storms are happening at wildly more frequent intervals than they have in the past centuries on which their "average expected frequency" is based. The Red River won't likely wait 500 years to rise that high again and may not wait even 50 years or 5 years. In Washington, DC, the Potomac River brought a 200-year flood in November 1996, then had *another* 200-year flood three weeks later. On both occasions, the volume of water flowing past (and through) the city expanded to about six times its normal volume—beating a record that had been set by Hurricane Agnes in 1972 and destroying a large section of the Chesapeake and Ohio Canal National Park. Yet, by the end of 1996, most Washingtonians were so numbed to climatic irregularities—or else memories had grown so short—that many were not even aware that it had been an unusual year.

Does it make sense to rebuild? The post-disaster interview has become almost a ritual, and the ritual response to having one's home or business destroyed by a natural disaster is *we won't be scared away*. It's heroic to rebuild. In *The Control of Nature*, John McPhee writes about a California family who built their dream house on a slope of the San Gabriel Mountains above the city of Glendale. Geologically, the San Gabriels are extremely unstable. After heavy seasonal rains, massive "debris-flows," which McPhee describes as having the consistency of hundreds of tons of fresh concrete, slide off the steep sides of these 10,000-foot mountains with regularity. Once some years ago, a big flow came down and smashed through the back of the Glendale family's dream house, filling it to the eaves within six minutes. The parents and their two children were buried alive and for a long time were presumed by rescue workers to be dead. But they survived by keeping their heads in the tiny air space remaining under the roof. Later, though it was made clear that their house had been directly in the path of recurring slides, and that another slide would come again in the future, the family resolutely decided to rebuild on the same spot.

Wherever natural disasters occur, that kind of response is seen—as it was in the Red River Valley. An article in *USA Today* focused on the residents' valiant stand against nature, under the title "Defying the Red River, Day After Day." The *Christian Science Monitor* waxed almost celebratory about such defiance in the article "In 500-Year Flood, Finer Things Surface." And Grand Forks mayor Pat Owens became a media favorite with her refusal to be discouraged. "I don't feel bad," she declared as she surveyed her wrecked city. "I don't feel bad for one reason, and that is that we will rebuild and come through this." But *is that really a good idea?* Rather than treating Owens' kind of determination sentimentally, reporters could raise questions not only about whether individual families or businesses should stay, but about whether *any* future construction should take place in likely flood zones. Or, indeed, in an age of accelerating climate change, whether whole communities should begin preparing to move.

The implications for fossil fuel consumption: If we know atmospheric carbon dioxide is spiking, and that climate scientists are in general agreement that this is disrupting weather and increasing flood risk the world over, then it's clear that assessing what happened in Grand Forks is not simply a matter of finding, as Mayor Owens puts it, "how vulnerable we are to nature." Rather, it means asking whether floods of this magnitude don't heighten the urgency of the carbon emissions issue. Can society afford to continue massive burning of coal to generate electric power, or gasoline to commute longer and longer distances?

The same year the Red River rose up, delegates of 160 nations were preparing for their climate convention in Kyoto to set "binding" limits on greenhouse gas emissions. The United States, despite its assumed post-Cold War role as Leading Nation, was doing no leading on this front. The US delegates, instead of preparing their country to rein in its giant coal and oil industries and shift its hugely disproportionate share of world energy consumption to technologies that don't emit greenhouse gases, were engaged in finding legalistic tricks to achieve the greenhouse equivalent to restructuring debt.

Those tricks, which were mostly smoke and mirrors, involved

improving the efficiencies of the current fossil fuel economy, instead of replacing it.[2]

An observer for the nongovernmental watchdog group Climate Action Network, observing the negotiations from the sidelines, wrote in his journal one night: "US representatives have held fast to loopholes that the entire fossil fuel industry could pass through, ample girths notwithstanding."

In the meantime, public officials were thoroughly distracted from the real need: to begin shifting the entire global energy economy away from the burning of coal, oil, and wood. They were repeatedly caught up in diversionary disputes about how much of the burden should be borne by industrial versus developing countries, or how much credit should be granted to polluters who promise to

2 ❧ For example, there was the scheme known as hot-air trading. Companies would be given emissions "allowances" and could sell their allowances to other companies. This would presumably provide an incentive for a coal-burning plant in Russia, say, to install cleaner-burning equipment so it would have left over allowance to sell to a plant in Illinois. But it would also mean that if the Illinois company were still making profits with a dirty plant, it could by buying the allowance be spared the capital cost of having to clean up its act. In reality, however, since the collapse of the Soviet Union had caused industrial production in the former Soviet countries in the late 1990s to fall well below the 1990 level on which the allowances were based, the Russian plant would likely have allowance to spare (and sell) without having to install cleaner equipment. In this scenario, neither plant would actually have to cut its emissions. The politicians would then be able to take credit for implementing a climate treaty without having incurred the wrath of the coal industries that were selling to the Illinois plant (and to thousands of other plants using the same trick), and meanwhile little or nothing would have been done to reduce global warming.

help trap more carbon in "sinks" by planting trees, which absorb CO_2 through their leaves. Delegates from New Zealand even wanted to get credit for tree plantations they'd planted in the *past* in order to allow their country larger amounts of coal or oil burning in the future. Exasperated scientists pointed out, to anyone who would listen, a sobering math: that even if the climate treaty as written were to be fully ratified and enforced, global carbon emissions would actually *increase*. The trading and the sinks were dangerous diversions. An enormous chasm was opening between the slow fiddling of climate politics and the furious reality of the climate itself.

These diversions were hauntingly similar to the tactics that had been used in the global population summit in Cairo, two years earlier, in which disputes about abortion pulled delegates' attention from the critical issue of how to slow population momentum. It seemed that a pattern was developing in public discussion of certain overarching global issues: whenever the big picture began to come into view, it was immediately obscured by a sudden, heavily financed dispute over an emotionally charged side issue. The larger issue was kept at a distance. Thus, during the summer preceding Kyoto, climate change continued to be treated by the media as an isolated topic—something for the science page, nothing to do with how or where people were living, or how they were preparing for the future. When the 500-year flood struck North Dakota and Minnesota, reporters made no connection between the rising of the Red River and the rising urgency of the energy source issue.

So, why were these issues—the causes, the chances of recurrence, the implications for what we do next—not discussed, either in the *Post* article or in most of the other media?

Part of the answer, of course, is political anxiety. The fossil fuel industries have such a grip on the US economy—and on the funding of election campaigns—that public officials rarely suggest that it's time to phase them out. Even the idea of cutting them back a *little*, as the climate treaty called for, is something US politicians dread to support. What the oil, coal, utility, and auto executives would like is a completely free market, in which not even the sky is the limit.

When the likely terms of the treaty became known in the summer of 1997, the US Senate voted 90-to-0 to oppose it unless the "developing" countries carried a substantial part of the burden. The senators' unanimity didn't reflect any strong feelings about the responsibilities of Botswana, Bolivia, or Bangladesh, however. The developing country issue provided an easy way for them to avoid becoming targets for political extermination by their largest contributors—the directors of Exxon, GM, Dupont, and the like. And, of course, it was not just the coal and oil industries that made up this sugar-daddy coalition to which no US senator could say no. It was also all the other satellite industries that are *dependent* on oil or coal: the electric utilities, car makers and dealers, and processors of petroleum-based lubricants, plastics, and even the asphalt used to pave the world's highways.

A few weeks after the Red River flood, Vice President Al Gore flew out to visit to its victims. It was a prime opportunity to point to a connection—between mounting weather disasters and global warming—that most Americans hadn't really begun to think about. The Kyoto climate conference was still in the future, and Gore could have seized the opportunity to galvanize the public and call attention to what was really at stake. But he didn't, just as the governors of North Dakota and Minnesota didn't, and supervisors of the flooded counties didn't, and the mayor of Grand Forks didn't. And if politicians avoid talking about an issue, at any level of government, the issue doesn't seem central to most people—it's pushed to the margins of consciousness.

Gore found himself trapped in an excruciating dilemma. On one hand, he and Bill Clinton had been elected in large part because of their championing of global free trade, which had brought them the support of multinational businesses that might otherwise have preferred a more conservative ticket. On the other hand, Gore had earned enthusiastic support from environmentalists, most of whom had heard the story of how a nearly fatal accident to his young son had aroused his determination to defend the global environment for those who come after us. During the election campaigns, conservatives had tried to portray Gore as an environmental "extremist" who would make trouble for business. Now, if he spoke out strongly on climate change, the conservatives would pounce on him with "we-told-you-so" zeal that could jeopardize his presidential hopes. If he didn't speak out, his environmental supporters would denounce him. During the year after Red River, he began speaking in guarded terms, focusing mainly on the relatively uncontroversial theme of the importance of the climate threat as a reason to improve the efficiency with which energy is used. In early August 1998, Gore noted publicly that the first half of the year had been the warmest on record. Sure enough, the conservative *National Review* immediately published a cover story depicting the vice president's face dripping like that of a criminal suspect under interrogation, with the headline "One-Man Heatwave: Al Gore's Bogus Global Warming Crusade."[3]

3 ❧ The article inside, by Jonathan Adler, was titled "Hot Air," followed by the statement "Global warming is not a threat to health or the economy. Plans to address it are." The month in which that article appeared was the 15th consecutive month the earth's average temperature had set a new all-time record. That same month, warming-driven weather disasters for the year to date reached $72 billion worldwide, breaking the old record of $60 billion with four months still to go.

For the rest of the world's opinion makers, however—its reporters, teachers, publishers, and policymakers—the silence in the late 1990s was not so much a matter of self-muzzling as of oblivious disconnection. With few exceptions, even the most highly educated people showed few signs of concern about the connections between the vast acceleration of human powers that supports their fast-track lives and the vast acceleration of change taking place in the climate and health of the planet.

Two months after the Red River flood, a similar disaster struck central Texas. The year before, that area along the lower Colorado River had experienced a severe drought. But in June 1997, some parts of it were flooded by water deep enough to completely immerse a four-story building. Governor George Bush Jr. was quick to request federal disaster aid. He made no mention of climate change, or of the possible role of oil.

It's clear enough why Bush kept quiet. Since oil has been a principal source of Texas wealth (and of Bush's personal wealth and standing), he could even less afford to speak out than Gore could. But that doesn't account for the obliviousness of the millions of voters who put those two political adversaries in office.[4]

4 ❧ That obliviousness prevails throughout the world, despite sporadic wake-up calls like that of the *World Scientists' Warning to Humanity*. A global survey conducted the year after the *Scientists' Warning*, by the George H. Gallup International Institute, asked people what they thought were the most important problems facing their nations. In no country did a majority of respondents mention environmental issues. In Russia, where the environment had been ravaged by Soviet chemical and nuclear dumping and other abuses, only 9 percent of the public cited environmental problems. In Brazil, for all its rainforest burning, only 2 percent did. Other totals included 9 percent in Germany, 3 percent in Great Britain, and 1 percent in Poland and Hungary.

❧

There are other forces, besides political fear, that have combined to veil awareness—to truncate our perceptions not only of aberrant weather disturbances like the Red River or Central Texas floods, but of surging flows of illegal migrants, the sudden appearances of previously unheard-of diseases, and the failures of once-healthy crops. They have not only veiled, but confused, our perceptions of the four megaphenomena by which we are now being so heavily affected with so little apparent awareness. Among the most powerful of those obfuscating forces are these:

- A creation of false extremes by corporate PR managers who issue "news releases" and "statements of concern" to media, leading reporters to shift their frames of reference on critical issues;
- A shrinking of vision resulting from the acceleration of change;
- A fragmentation of knowledge resulting from specialization in virtually all areas of human knowledge;
- An informational black hole resulting from an expanding global shadow economy.

FALSE EXTREMES

One of the guiding ideals of journalism is balanced reporting, which is achieved by making sure you have presented both sides of an issue. The origin of this view is innocent enough; it's based on the assumption that on most issues of interest to the public, there's a wide spectrum of opinion. The democratic ideal demands that all sides be heard. A reporter's concept of objectivity requires that if someone on the left expresses an opinion for a story, someone on the right should also be heard in order to keep the story balanced.

This premise seems fair enough, but it contains a couple of dangerous flaws. Not all issues are matters on which there may *be* a real spectrum of views. There's no longer any debate worth reporting about whether cigarette smoke causes lung cancer, or whether unshielded UV light burns human skin—*or* whether phasing out the use of coal and oil for energy would reduce the risk of a dangerously destabilized climate. However, each time new information comes out to further embarrass an industry whose products have been widely incriminated, we've seen that that industry can still sow confusion—and thwart action against it—by aggressively acting as though the issue were still very much alive. Often, a reporter can be persuaded to offer the embattled industry "equal time" to vehemently assert that the new information is untrue. The reporter is persuaded on the grounds that this constitutes balanced reporting. Obviously, most reporters know well enough that assertion is not fact, but they are often deceived when the assertion is presented—by people presumed to be experts, or by PR people acting like fellow reporters—in a form that makes the assertion look like fact.

The first major use of this technique in response to a key environmental warning occurred after the publication of Rachel Carson's *Silent Spring* in 1962. Today, that book is widely recognized for having launched the environmental movement in the United States. When it first came out, however, its warnings about chemical pollution were blunted by a massive countercampaign by the chemical industry. The National Agricultural Chemical Association hired a PR expert, E. Bruce Harrison, as manager of environmental information. As recounted by authors John Stauber and Sheldon Rampton, the campaign used "scientific misinformation, front groups, extensive mailings to the media and opinion leaders, and recruitment of doctors and scientists as 'objective' third-party defenders of agrochemicals." The Monsanto Corporation published *The Desolate Year*, a retort to Carson suggesting that failure to use pesticides might cause a plague of insects that would devastate America.

Over the next three decades, such counterinformation campaigns became commonplace, and the techniques grew more sophisticated. During the climate treaty negotiations in 1997, news organizations were inundated with what appeared to be photocopies of articles in scientific journals concluding that climate change is not a problem. Those articles were not what they appeared to be.

Because large changes are always alarming or unsettling to at least a portion of the population, any new evidence that such changes are coming will find a certain segment of the population wanting to believe it isn't true. That means there's always a ready market for denial. If the discovery is as alarming as the consequences of the four megaspikes are, the market for denial may be very large. The reporter who ventures to include the dissenting view in the interests of balance often finds his or her efforts seeming to be validated by this market: the people who are in denial buy copies, write applauding letters to the editor, and provoke exasperated reactions from those who are being contradicted. This creates a lively controversy, which many editors and publishers relish, so the reporter is further encouraged. Readers who previously had no reason to question the facts now become uncertain. A spectrum of opinion is created where none previously existed. For example, an editorial in a major Ohio newspaper, in the summer of 1996, had this to say about global warming:

> On one side, believers such as Undersecretary of State Tim Wirth want to reduce US carbon-dioxide emissions by 10 percent under a grand international treaty that would hamstring American industry while letting developing nations such as China and India grow unabated. On the other side, numerous scientists say more evidence is needed before attempting anything so drastic.

The implication that Wirth is alone at one end, opposing "numerous" scientists at the other, turned the truth on its head. In fact, the consensus of scientists was quite in agreement with Wirth, and even considered his position rather modest. In the meantime, however, decisions on how to respond to the apparent "controversy" are delayed, as policymakers try to sort out their mounting confusion. And the world loses more ground as it loses valuable time.

That's the mechanism at its simplest, but a more sophisticated version is becoming prevalent. The spectrum of opinion is often confused with (or assumed to be) a sort of bell curve of public belief, with the center representing the mainstream. Implicit in this is the idea that the mainstream view is the one that has the greatest weight of evidence supporting it—and that the extremes represent only marginally supportable views. It's a hubris of democracy—an assumption that the majority can't really be wrong.

That idea has been deftly exploited by those who believe their business dominance will be threatened by any major international mobilization to stop the carbon-gas spike. When the world's climate scientists issued their first IPCC report on the likely consequences of that spike, the fossil fuel PR departments went to work on repositioning the IPCC from the center to the extreme, by finding a contrarian view to take the opposite extreme—thereby placing themselves in the center. There are always a few people who are willing to play that contrarian role, if the rewards are sufficient—whether those rewards be notoriety, generous speaking fees, or "funding for research."

In his book *The Heat Is On: The High Stakes Battle over Earth's Threatened Climate,* journalist Ross Gelbspan quotes a memo circulated by coal executives describing the aim of the campaign to "reposition global warming as theory rather than fact." The campaign even identified three contrarian "experts" who could be counted on to debate the IPCC scientists. The call for debate immediately created the impression that the question of climate change was actually a matter of soul-searching controversy. In their directive, the coal executives stipulated that the views of their hand-picked contrarians be "placed in broadcast appearances, op-ed pages, and newspaper interviews."

The debating techniques used by these contrarians rested heavily on the fact that scientists, by the nature of their disciplines, are rarely willing to say they are absolutely certain. The public, accustomed to being given assurances of unqualified certainty by everyone from weight loss consultants to talk radio commentators and criminal defense attorneys, easily assumes that any expression of uncertainty must betray a truly weak case. Yet, the kind of uncertainty expressed by the IPCC scientists had to do largely with such questions as *how fast* climate is changing as a result of human actions, and *how much* storm intensities will rise. But to a public not following the details, this was easily blurred into an impression of uncertainty about *whether* the climate will really change.

A typical example of such fabricated controversy was exhibited in a debate held by the World Future Society and published in the March/April 1997 issue of the society's journal, *The Futurist,* under the title "The Global Environment: Megaproblem or Not?"

The title alone implied that there is a spectrum of views on whether there is a serious problem. Representing the contrarian view was a University of Maryland economist, Julian Simon—one of the experts who had been recruited to carry the banner of the fossil fuel industry. In the debate, Simon led off with a statement that "our air and water have been getting cleaner rather than dirtier in the past few decades." As evidence, he cited the amount of smoke in the air in London, which has decreased greatly since Victorian times. "Smoke level has been the most important pollutant in the air for hundreds of years," he told his audience. "This is the dirty black stuff that gets in your lungs and kills you in London or Pennsylvania." And of course, the kind of smoke Simon was describing—the soot that soiled London and Pittsburgh a century ago—*has* diminished. But global warming has nothing to do with soot; it is a problem of CO_2, which is invisible and is spiking. Simon never mentioned CO_2—apparently assuming that for his listeners, out-of-sight is out-of-mind.

Climate scientists dismissed Julian Simon as an unprincipled publicity seeker, but ironically his message about smoke created enough rhetorical smoke in the 1990s to blur perceptions among the public at large and to create willing adherents among those—including many corporate executives—who do not want to believe something radical is happening to their world.

That set the stage for an even more subtle campaign, in which fossil fuel executives persuade fellow senior executives in allied industries that because there's a wide spectrum of opinion about climate change, it would be premature for government to take action. The fellow executives, needing little encouragement to oppose further government constraints on business, readily sign on to calls for caution on the policy front. The outcome can cripple public action, as was demonstrated in the summer of 1997—just a few weeks after the Red River flood. Just as the Earth Summit delegates assembled in New York that June, to agree on the levels of binding reductions each country would commit to at the Kyoto Climate Convention in December, a two-page newspaper ad appeared in the *Washington Post*—undersigned by a long list of major corporate executives known as The Business Round Table. The ad, in giant type surrounding a photo of planet Earth, announced:

WITH A BALANCED APPROACH
WE'RE COMMITTED TO A HEALTHY ENVIRONMENT
AND A HEALTHY ECONOMY

It was signed by the chief executives of 160 major corporations (a neat counterweight to the 160 nations represented at the Earth Summit), including such fossil fuel-dependent giants as American Electric Power, AMOCO, ARCO, BF Goodrich, Chrysler, Dow Chemical, Dupont, Ford Motor, General Electric, General Motors, Goodyear, Mobil, Occidental Petroleum, Pennzoil, Phillips Petroleum, Shell Oil, and Texaco. It called attention to the ongoing climate discussions and, to anyone reading it casually, it may have seemed to say that these companies were deeply concerned about our planet's environment and wanted to do "the responsible thing." Only in the small type—the text that congressional aides and White House advisors would be sure to read—was the real message:

> Later this year, the Clinton/Gore administration must decide whether to sign an international climate treaty designed to reduce the potential dangers of global warming. It may be the most important economic decision of this century and the next as well. So far there has been little public debate on the treaty even though it

> could have a dramatic effect on the way we live and work here in the United States. . . .
>
> At the Business Roundtable, we have a responsibility to ensure future generations will have a healthy economy and enjoy an ever-increasing standard of living. We have an equal responsibility to protect our environment for future generations. We believe it's possible to do both with a balanced approach.
>
> But a balanced approach is only possible with careful study . . . and extensive public debate. We strongly urge the Clinton/Gore administration not to rush to policy commitments . . . until the environmental benefits and economic consequences of the treaty proposals have been thoroughly analyzed.

To the policy people in Washington, the subtext was unequivocal: Don't commit US industries to anything. The signers represented a large fraction of US Gross National Product (and Gross World Product) and an even larger portion of campaign financing for both US political parties. In their call for "debate," they reinforced their carefully cultivated idea that there is a spectrum of opinion about whether climate change is really a serious problem and then nimbly assumed the center position on that spectrum—the position of open-minded citizens merely interested in hearing more and not rushing to judgment. Through their ad, it was the oil executives and their colleagues, not the climate scientists, who appeared to represent the most reasonable, or mainstream, position. A few days later, Bill Clinton did exactly what the ad suggested he should do: in a wordy speech to the United Nations, he committed his country to nothing.

Each of the other megaphenomena, likewise, has borne the brunt of managed marginalization by well-financed corporate PR campaigns. Perceptions of population growth, unsustainable consumption, and rising rates of extinction have been distorted by a coalition of interests that give unyielding primacy to humankind over all other life-forms. This coalition includes pro-life groups that have opposed funding for family planning because they believe it encourages abortions. It includes so-called technological optimists, such as the publishers of the Lyndon Larouche-funded magazine, *21st Century Technology*, which espouses the wonders of escalating human powers over the planet while dismissing reports of the ozone hole and other signs of human-caused damage as outright "hoaxes." And finally, this coalition includes pro-free trade enthusiasts, such as the publishers of the magazine *World Trade*, who have embraced an equally unquestioned faith in the rich rewards of unfettered capitalism.[5]

Disparate as these factions are, they share two general features: they put the burden of solving the world's problems in *others'* hands (either God's or the inventors and entrepreneurs of future technologies), and they base their belief on faith. This appeal to faith provides the basis for the fabricated spectrum in each case, as science is pushed to a false extreme at one end of the spectrum. The other extreme is easily identified with utopian visions of the fringe New Age or Aging Hippie movements, leaving the sober-minded Business Roundtable and techno-optimist views to occupy the center. The New Agers and Aging Hippies don't participate in the debate, since they are presumed to be

5 ❧ An article in the March 1997 *World Trade* began, "Grab your ax, oil your chain saw and head for the tall timber. Canada's forestry sector offers US firms a thick woodland of selling and alliance opportunities. . . . Canada has more than 12,000 forestry operations, ranging from remote logging camps to mammoth wood manufacturing plants, every one of them needing custom gear and supplies to keep the logs rolling, the saws humming, the mills churning."

content to wait for the Age of Aquarius or Universal Synchronicity to sweep away the present order, undoubtedly throwing us all back to the Stone Age. That leaves the scientists (other extreme) and the businessmen (center) to talk it out. The resulting "debate" concerns the extent to which the problems of population and consumption will be solved by what the overanxious scientists say we must do now, versus *what we shall discover in time*—whether through God's will, human ingenuity, or the market. And in the United States, at least, this pro-life, pro-trade, techno-optimist coalition has succeeded in getting many policymakers to believe that it represents the mainstream—the voice of reason.

In the case of the extinctions spike, it is the timber and paper industries that have performed the most effective managed pushing of science to the extreme. These industries own or lease huge tree plantations, which have increasingly replaced natural forest ecosystems and destroyed vast areas of habitat. By displacing habitat, they have also destroyed much of the biodiversity that habitat shelters. These industries argue that in the United States, the amount of forest cover has actually increased in the past century. This, like the Julian Simon "smoke" story, is technically true and absurdly misleading. It fails to mention that much of the reforestation is either weak secondary growth or monocrop plantation that is leaving the land genetically depleted—another apparent reason for the rising numbers of extinctions. This argument also fails to note that the increase in the United States is an exception to the global trend. Elsewhere in the world—in the vast boreal forests of Russia and Canada and the great tropical forests of Indonesia, Papua New Guinea, Malaysia, Mexico, Suriname, and Brazil—the tree cover is disappearing at free-fall speed.

CONTRACTING VISION

The second force impeding our ability to see critical megaphenomena is a highly disorienting *acceleration* of change. Imagine a car losing its brakes as it descends a winding mountain road; as it gains speed, the curves become harder to handle. The changes we're encountering are like those curves; with each passing moment we have less time to anticipate—to plan for what lies ahead. In effect, it becomes harder to see into the future with each passing year or month.

That may seem counterintuitive, given the communications and forecasting technology we have at our disposal. We can look over our shoulders at the great visionaries of the past—Leonardo da Vinci, Charles Darwin, or Jules Verne—and reassure ourselves that we now have forecasting capabilities (such as computerized demographic or climate models) that even they never dreamed of. However, there are good reasons to believe that we probably have far *less* capacity to envision our next century of future than they had to envision theirs.

To begin with, a person living sometime in the one thousand centuries before societies began keeping historical records probably had very limited vision—or vulnerability to change—from a *geographical* standpoint. From anthropological research, we can infer that he or she probably never traveled far (even the great migrations from Africa or across the Bering Sea land bridge probably took many generations) and had no knowledge of distant peoples or lands. Throughout those long millennia, most people likely experienced less change in a whole lifetime than we could undergo in a few months or even days. In the slow passing of a skill in stone-chipping from father to son to grandson, perhaps the shape of a spear-point changed. On the climate front, perhaps there was a slight warming or cooling. When so little changed, it was not so hard to plan for what lay ahead. There may have been an occasional raid or war, but never with the kinds of intercontinental impacts the wars of the twentieth century have had. In the twentieth century, wars went, for the first time, from regional to world scale. They brought, also for the first time, the possibility of global annihilation; they killed more people (about 110 million) than the wars of the previous twenty centuries combined; and they shifted the burden of suffering and death from soldiers to civilians.

Equally enormous accelerations have occurred, in the twentieth century, in communications, transportation, medicine, and resource extraction—the mining of materials and energy to sustain all those other accelerations. Whereas the changes that took place in human life over a 100-year span during millennia past might be summarized in a page, what happened in the twentieth century would—and does—fill thousands of books.

Now, consider that the abrupt acceleration that took our transportation technology from the horse-drawn carriage to interplanetary flight in a single century is continuing exponentially. We have heard ad nauseum about the expansion of computer capacity, but this kind of expansion is also happening on a multitude of other fronts we have heard much less about—and in some cases may *never* hear about. It's significant that even as NASA scientists were preparing the Sojourner exploration of Mars and the Casini probe to Venus and Saturn in 1997, other scientists were probing realms too small in size for even electron microscopes to see. And those are not the only directions exploration is going; there are the frontiers of the human genome project, of our evolutionary past, of interspecies communication, of outer space, and of the possibility of an evolutionary leap.

Unfortunately, our ability to plan for the radically expanding world we are now entering depends on having stable frames of reference, or sets of known underlying conditions from which to launch credible projections into the future. The projections are "if-then" statements, not prophesies; they tell us that *if* certain existing or foreseeable underlying conditions prevail 10 or 20 years from now, *then* here is what will probably happen. Of course, underlying conditions will rarely remain exactly the same over time, but in the past they have changed slowly enough to provide reasonable continuity. But if those underlying conditions are *themselves* changing faster than we can keep track of, it's not possible to construct a dependable projection. If the "if" is highly uncertain, then the "then" is unknowable. If we can't see the curves coming, we have little hope of staying on any preplanned course.

One of the most critical areas in which this contracting of vision now endangers our ability to plan is in the production of food. Food supply forecasts are of critical importance to governments, whose best-laid plans can be thrown to the high winds if hit by famine—as has happened during the past century in the Ukraine, China, India, Bangladesh, Ethiopia, Eritrea, Somalia, Sudan, and North Korea. Over the past half-century, global food forecasts have been made mainly by two organizations: the World Bank and the UN's Food and Agriculture Organization (FAO). In that half-century, a seemingly reliable pattern emerged: as population grew, so did yields of grain—the basic food staple. If more fertilizer was added and more dryland was irrigated, and if new varieties were developed so that more photosynthetic energy was concentrated in the seed and less in the stalk, then more grain—either for direct consumption or for feed to produce meat—could be harvested. For four decades, the formula worked, and the forecasters assumed it would continue to work well into the twenty-first century.

In the 1990s, however, the underlying conditions began shifting rapidly—in ways the forecasters failed to take into account. One of those conditions was the assumed continuing availability of land for crops. In the past, when population grew, it was always possible to clear more land for farms. Planners probably never dreamed the day might come when there would be *less* land available for farming than before.

Yet, this is what happened in the last years of the twentieth century. It happened gradually, for several seemingly disconnected reasons. One was the rapid expansion of cities, all over the world. The locations of most centers of population were originally chosen because the land was good for farming, so as cities have exploded in size they have spread over fertile land at a disproportionately rapid rate. The rivers that provided good soil and water also provided means of transport that became important to trade, so the tendency of people to settle near water—and therefore on good farmland—was reinforced. In the past quarter-century, millions of acres of the world's richest farmland have been covered with new housing developments, industrial parks, and pavement.

In China, for example, the superheated economic growth of the 1990s produced an intensifying competition between farmers and developers for the land around Beijing. The farmers lost, and thousands of them were kicked out. In Vietnam, the same thing happened on the outskirts of Hanoi. The government enacted a ban on industrial development on land used for growing rice, but four months after the ban was passed, an exception was granted to Ford Motor company and some of its satellites to build factories on 6,310 hectares (about 15,000 acres) of prime farmland near Hanoi. The cars turned out by the factory, of course, then stimulated the demand for more roads and parking lots, which brought the paving over of yet more Vietnamese farmland.

Worldwide, the amount of land planted in grains—the main crops on every continent—reached a peak in 1981, after 10,000 years of expansion, and has declined by 5 percent since then. The forecasters at the FAO and World Bank neither foresaw this shrinkage of land nor, when it began to happen faster, thought to add it up. They continued to make the same assumptions they always had. Unfortunately, if the "if" in an if-then forecast is blown away by unexpected change, the "then" can be tragically wrong. And in the case of the world's food supply forecasts, it is not only the assumptions about how much land is available that have been blown away, but other key assumptions as well. The strategies of adding more fertilizer to increase yields, or of developing more seed-heavy varieties, or of tapping more water for irrigation, all have reached points of diminishing returns.

Fortunately, a few vigilant experts on these various strategies—on freshwater supply, fertilizer applications, plant physiology, etc.—took notice of the abrupt changes in their respective areas, and a few ecologically minded analysts such as the population experts Lester R. Brown and Paul Ehrlich began putting together the bigger picture. In 1994, an article in *World Watch*, "Who Will Feed China?" by Brown, systematically examined each of the conditions assumed by the World Bank and FAO projections in their standard forecasting formula, on which China's (and the world's) agricultural planners were depending. The analysis confirmed that all of those conditions had been blown—that China will not have anywhere near enough productive capacity to feed its surging population in the twenty-first century, and that, for the same reasons, the rest of the world's countries *combined* won't have enough grain export capacity to make up the difference. The story was picked up by global media, and the Chinese government called an extraordinary press conference to denounce it. Three weeks later, though, the Beijing government reversed itself and conceded that Brown's assessment was right. China had a giant problem.

In any case, tracking changes in baseline assumptions for forecasting isn't the hardest part of keeping up with the accelerating pace of change. Still harder is anticipating the emergence of entirely *new* factors affecting what will happen. Again, in trying to forecast the future of the food supply, consider that the spike of global commerce has brought a huge increase in the movement of nonnative species of plants, animals, and microorganisms over the planet. Some of them have ravaging effects on food crops. Until about four centuries ago, this movement was minimal. Few people traveled far, and those who did traveled by modes so slow that few exotic organisms could hitchhike in the traveler's baggage and survive the trip undetected. But among the rising spikes-within-the-spike of global consumption, we have seen surging growth in jet-transported international tourism, migration, and trade—including black-market trade.

Tourism is the fastest-growing industry on Earth. Refugee and migrant flows reached their highest levels in history in the 1990s. And the volume of officially reported world trade more than tripled between 1970 and 1995. Unofficial trade has apparently grown even faster. With marine organisms carried in the ballast water of giant ships, viruses in the lungs of millions of tourists, and exotic animals or plants in the jet planes of high-volume wildlife traders, the biosphere is being subjected to unprecedented biological mixing. Among the mounting "bioinvasions"—introductions of organisms to ecosystems where they did not evolve and where they often have no natural predators to keep them in check—are hundreds of new agricultural pests. As my colleague Chris Bright has documented in *Life Out of Bounds: Bioinvasion in a Borderless World*, crops come under assault by new pests every year, and those new pests constitute a factor not accounted for in the traditional formula for forecasting food supply. Meanwhile, the use of pesticides offers no lasting solution, as insects now develop resistance to insecticides as fast as chemists can develop new ones—and the health effects of pesticides put into mass use with little or no field testing have turned into a tragic human burden.[6]

What happened in the misforecasting of world food capacity is likely to occur in a wide range of other areas, as once-stable conditions become increasingly unsettled. Where we once moved toward the future on fairly firm ground, we now move on treacherously shifting sands. The most critical of those other areas come under the umbrella of "human security"—meaning protection not only from starvation, but from *all* physical threats. If technology originally grew out of a desire to magnify the capabilities of the human body and brain, *security* became the protection of the body from the misuse of that same magnified capability.

Until the late twentieth century, the conditions determining security—like those determining food supply—were based on relatively stable sets of conditions. Perhaps the most funda-

6 ❧ In 1990, Cary Fowler and Pat Moony reported in their book *Shattering: Food, Politics, and the Loss of Genetic Diversity,* that "over 400 species of pests have now developed resistance to the chemicals that once destroyed them." Six years later, researcher Gary Gardner of the Worldwatch Institute reported that that number had risen to 900 species—with 17 crop-eating insects showing at least some resistance to *all* major classes of pesticides.

mental of those conditions was a world made up of sovereign nation-states, which had well-guarded borders and which controlled all important activity within those borders. Armies were designed and equipped to defend those borders, and intelligence services were trained to assess the capabilities of opposing armies. But in many respects, the nation-state is becoming obsolescent. The gates have opened so wide to human (and other) traffic that national borders are unlikely to ever again delineate militarily defendable territories as they once did. And peoples have become so mobile and intermixed that the prospect of defending a people by deploying tanks along the borders of their country is increasingly doubtful.

In the conflicts of recent years in Bosnia, Rwanda, Cambodia, Colombia, or the former Zaire, the adversaries were not defenders of territories, but neighbors within the *same* territories. Of the approximately 40 wars fought in the world in the 1990s, 39 were civil wars or guerrilla wars *within* ostensible national boundaries. Those wars were no longer about the geopolitical issues that drove the planning and funding of the US or Soviet armies, the CIA, or the KGB for so many years. Within a very short time, the spiking of human population and consumption has thrown into our path a whole new set of security threats, and most of today's wars are either triggered or exacerbated by these threats. They include:

- *Looting of resources by transnational corporations.* For example, there are the timber operations that are clear-cutting vast areas of Russian forest, or the mining operations that are taking copper and gold out of Indonesia with virtually no compensation to the indigenous people whose ancestral land has been seized, or the extraction of genetic resources from tropical forests by US or European pharmaceutical companies, which then claim exclusive world rights to the material as if they had invented it.

- *The emergence of new and resurgent diseases.* More than half a century after the discovery of antibiotics, infectious diseases are again on the rise. Tuberculosis, cholera, malaria, diphtheria, sleeping sickness, river blindness, and schistosomiasis now infect more than 2 billion people worldwide. One of every three people in the world carries the TB bacterium. And now, new infectious microbes have appeared that are resistant to all known antibiotics.
- *Use of weapons of mass destruction by groups or individuals not allied with any government.* By some measures, the risks of nuclear, chemical, or biological attack have risen rather than receded, but they're not the kinds of risks for which traditional military forces provide any real protection, because these new enemies have no geographical bases toward which missiles can be aimed, or fleets deployed.
- *Rising refugee and migrant flows.* The number of people who have been forced out of their homes or homelands has risen sharply throughout the last quarter of the twentieth century. The annual flow of refugees alone has climbed from about 2 million in the early 1970s to more than 25 million in the 1990s, according to data compiled by the UN High Commissioner for Refugees. If people who have gone into flight within their own borders are included, the number rises to more than 50 million. If economic migrants—legal and illegal—are included, the number rises above 100 million. The world's dispossessed or discontented have become a tide that now threatens to overwhelm the places toward which they flee.

These threats, and others, now greatly overshadow the threats of organized military invasion for which traditional security forces were designed. Intelligence services and security analysts, still fixated on the kinds of invasion they were trained to deal with, have been ambushed. Their countries have *already* been invaded, and in some cases their native cultures have already been gutted. It has happened in a flash, and the basic reason the planners haven't seen it is that it is unlike anything they expected. Like the World Bank grain-counters, they have been overlooking key indicators they assumed to be stable, but which have suddenly begun to shift. While military authorities have been counting troops and warheads, they have not been counting the number of people drinking contaminated water, or the number of people earning less than $1 per day—large numbers of whom will soon be coming their way.[7] Planners continue to pour funding into old problems while missing the new, and the result is that for them, visibility has dropped almost to zero.

FRAGMENTATION OF KNOWLEDGE BY SPECIALIZATION

The third force interfering with our ability to see whole pictures—to see complex chains of cause-and-effect, and not just immediate effects—is our increasing reliance on specialists as the

7 ❧ The United Nations Development Programme (UNDP) reported in 1998 that the number of people trying to survive on $1 per day or less had reached about 1.3 billion—more than the total populations of the United States and Europe combined. And according to Worldwatch Institute researcher Anne Platt McGinn, the number of people for whom safe water is now a "life-or-death issue" has reached about 1.2 billion—of whom 25 million are dying of waterborne diseases each year.

principal sources of knowledge. Of course, specialization has had great value to modern society; one can board an airplane, sit back with a drink, and enjoy the product of decades of aeronautic engineering research without having to understand any of it. But that also means we can enjoy the amenities of our lives without having any inkling of what chain of events brought them about—or of what it may have cost to produce them. After all, in an age of information explosion, the sum total of human knowledge—of what there is to see and to plan for—is expanding far faster than even the most brilliant people are capable of keeping pace with. One unarguable result: with every passing year, each of us knows a *smaller* fraction of what there is to know than we knew the year before.

The expert view, as it probes deeper, also becomes narrower. It penetrates further and further into less and less of what is visible to everyone else. The danger, for us, is vividly described in a 1997 article by Timothy Ferris in *The New Yorker* magazine, on the probability (or improbability) that our planet will one day experience an apocalyptic collision with a large asteroid or comet. Of course, small objects hit Earth every day, but the worry is about what will happen if a chunk of ice or rock the size of, say, the Rock of Gibraltar should strike our globe. Scientists say it would raise a tsunami—an oceanic wave—as high as a 70-story building, which would sweep such cities as New York, Miami, and Dhaka to oblivion. Its explosion, meanwhile, would pelt the planet with enough fireballs to set the planet ablaze and consume almost everything that was not drowned. The fire would send up enough smoke to plunge Earth into a flood-covered darkness that would last considerably longer than 40 days and 40 nights. And that would kill off most photosynthesis-dependent life, which would, via the food chain on which we all depend, eliminate nearly *all* life, including our own. One might consider it a matter of considerable curiosity to know just how probable this scenario is.

That depends, in part, on how many such rocks are at large in our solar system. Ferris reports that there are about 2 thousand large asteroids—ones "the size of a small town"—that pursue orbits bringing them close to Earth. It would be possible for astronomers to count them and keep track of where they are. A collision could be predicted days or perhaps months in advance, for those who consider knowledge for its own sake to have inherent value—which is to say, even if there were not much practical use one could make of it. So, you would think that scientists, who are among those of us who do value knowledge for its own sake, would be particularly curious about where these objects are. But to find out where they are would require studying the whole sky, and this is not what professional astronomers do. Astronomers are specialists; they pick what would be, to your eye, an extremely tiny piece of the sky—and, in football parlance, "go deep." And from *their* perspective, this kind of extreme specialization makes sense.

If you hold a grain of sand at arm's length, they might explain, you would cover a piece of sky that, seen through the Hubbell space telescope, reveals a region of the universe that contains perhaps a hundred galaxies, each galaxy containing perhaps a million suns. In other words, within that grain-of-sand's worth of sky would be found uncountable reaches of universe larger than the whole sky you see with your naked eye. What the professional astronomer—the specialist in that region—sees is, conceptually, the equivalent of viewing Vincent Van Gogh's *Starry Night* magnified a trillion times its original, nineteenth-century magnificence. To someone who has grasped the scale of that, our own minuscule solar system might seem too shallow to be worth the time of day. And so, as

Ferris points out in his article, if a doomsday object were ever to approach, it would most likely be discovered first by an amateur—perhaps a kid with a backyard telescope.

What happens to astronomers also happens to specialists in a thousand other fields—including those in the fields of economic, demographic, climatic, and ecological forecasting. In each, there are numerous subspecialties in which one must be a particular kind of expert to understand much at all. If you are such an expert and you discover something curious, there's a good chance that only your colleagues in the field can really grasp it. Most experts no longer try to keep in contact with the rest of us at all; they are like the motes in an explosion of understanding being carried outward from the center. Think of the center as the common ground of those of us who are still close enough to each other to be able to integrate our collective knowledge and make it work as a system; it is our cultural and ecological integrity. The way expertise is exploding, the center can't much longer hold.

The accelerating movement outward, or away from commonality, creates in each of these fields an illusion of independence, as though the field were governed only by its own gravity. That may make things appear more orderly than they will one day turn out to be. Within each field, the experts plot their trajectories according to their own rules—the rules of the field. And, within the self-contained world of their expertise, what they see has a beguiling logic. As long as it doesn't have to be reconciled with the logic of other fields, this internal logic may seem to make sense. The more highly developed a specialty, the more self-contained its logic.

An *architect*, for example, has a highly developed knowledge of the internal systems of a building, which in some ways—in its integration of energy use, heating, cooling, circulation of air and water, wiring, and disposal of waste—is a microcosm of the larger systems of the natural world. The architect will also have a fairly accurate understanding of the cost the client must pay for the materials used to construct and run that building—yet may know nothing about the larger cost to the world of extracting those materials. The architect can specify redwood siding, for instance, without ever consciously participating in the decimation of ancient redwood forests and of the great genetic wealth they contain.

A *pest-control technician,* to take another example, may be able to provide expert appraisal of whether the ants in your non-redwood beams are carpenter ants which will destroy the beams, or not. But the technician, to qualify for this highly specialized job, is not required to know anything—and indeed, almost certainly does *not* know anything—about the probability that the "product" he is spraying (that's the word he'll use) into all the cracks and crevices of your home will increase your daughter's risk of cervical cancer. "I'll need to put more product up there," he might say. So specific is his job that he may not even know what the chemical is. That job is for someone else.

An *automotive engineer* can explain why the new car engines are more fuel efficient than the ones we had decades ago and may even take pride in knowing that these more efficient engines have helped to clear cities of smog. But for the engineer to see the *real* problem with energy consumption would require stepping back and looking at a much broader picture, like the astronomer pulling back from his pinpoint focus to scan the whole sky. Because the real problem isn't in engine design at all, but in *urban* design. The engineer's highly specialized education is unlikely ever to have suggested that the fundamental problem may be the way we spread homes and workplaces so far apart that millions of people have to use cars, rather than trains, bicycles, walking, or even staying put for a good part of each week, to live their day-to-day lives.

It is in the arcane world of the economist, however, that the internal logic of a specialized discipline has become most insular. It is the dogmas of economics, rather than the laws of nature, which govern the globalism that now has us in its thrall. The most central of those dogmas is a belief in the logic of limitless growth. To economic theorists, growth *could* mean increasing wealth—or quality of life—without increasing energy consumption or pollution to unsustainable levels. It could be done by increasing the energy efficiency of products, by redesigning communities, by encouraging more sharing of underutilized capacity (do ten adjacent townhouses need ten power lawn mowers?), and by a number of other well-proven strategies. But most economists don't bother to make this clear to the political and business leaders for whom they work. So, to most people, economic growth just means more *consumption*. The spike continues to rise. And most economists, who aren't required to balance global accounts, go along with the illusion. As long as societies don't require them to reconcile their view of the expanding economy with the views now seen by experts on human population, or on the human carrying capacity of a single planet the size of ours, their pitch for limitless growth goes unchallenged.[8]

Economists learned long ago that they did not have to account for all the social or environmental costs of doing business—that

8 ❧ In *The Stork and the Plow,* Paul Ehrlich and Gretchen Daily write that the obliviousness of economists to the realities of planetary limits "traces primarily to a lack of basic education in the physical and natural sciences. This was illustrated in a 1987 survey of the opinions of graduate students in economics on the importance of other fields to their development as economists. The lowest score was given to physics. Only 2 percent of the students considered it very important . . . whereas 64 percent rated it entirely unimportant."

they could ignore any costs inflicted on those who aren't parties to the transaction, or don't speak the language, or aren't present when the deal is made. In centuries past, they—or the advisors who were their predecessors—learned that they could disregard the depletion of resources in distant colonies. More recently, they found they could disregard the pollution-caused medical costs that will be paid by individuals who have no way of knowing who was responsible—or, indeed, whether anyone was responsible. Fortunes have been made by inflicting costs that could be subsequently explained away—the way catastrophic storms and floods are so often explained—as "God's will."

Most economists, of course, don't work for humanity; they are hired by individual communities, countries, or corporations. Because the rules don't require balancing accounts for Earth as a whole, an economist can help his own country or county or company to gain by allowing another to lose. Specialization protects the rules that allow this incomplete accounting in several ways. First, it's hard for a family buying Nike sport shoes in Bucks County, Pennsylvania or Osaka, Japan to be aware that those $100 shoes were made by workers laboring for 30 cents an hour under near-slave conditions at the Hardaya Aneka shoe factory in Indonesia, because the disconnection of economics from human rights doesn't require that economists inform us of the full costs of our lifestyles. Second, when people who see across the boundaries of disciplinary fiefdoms *do* speak up—when they do try to alert the public to the full implications of their activities, appetites, and purchases—they are often pushed aside by heavily financed marginalization. In short, the rules of specialization still succeed in keeping us in the dark for the same reason that a pyramid scheme succeeds: because at least for a while, each new customer is unaware of the overall picture.

THE INFORMATIONAL BLACK HOLE

Finally, our ability to plan for epochal changes—changes that will require exponentially faster and more intelligent adaptations than we have ever achieved in the past—is being constricted not only by the fragmentation and sometimes deliberate marginalization of information, but by a large-scale *disappearance* of information that may be important to our survival.

The spike of world commerce and consumption has so dominated the discourse of public business in recent years that it has distracted us from the reality that large portions of activity are not integrated into the global system—not accounted for in the statistics of Gross World Product, or the World Trade Organization, or the global watchdog organizations that monitor climate-altering gases or biological diversity loss. That's not entirely new, of course; before the industrial age, it might be argued that *most* human activity was not accountable in the modern sense. People did not have Visa account numbers, Social Security numbers, business tax IDs, or numbered street addresses. But now that we assume most do, it has become a matter of rising concern to governments that large numbers of people continue to do what they do under cover of what has been referred to as the shadow economy.

That term is used broadly, to include sectors ranging from expanding organized crime to what economists call—and incidentally often marginalize as—the informal sector. Whatever the reason may be that a particular group of people is keeping (or in some cases is kept) out of the public view, that lack of visibility can have consequences of a kind that simply didn't exist in the pre-globalization era. One consequence is that activities which threaten human security on a global scale may go unmonitored and unregulated by any government. Black-market commerce in nuclear, chemical, or biological weapons, for example, is evidently being conducted at growing risk to the global public. Another consequence is that people kept out of public view may be deprived of public protections—as are the thousands of young girls kept as slave-prostitutes in Thailand, or the millions of illegal migrants deprived of clean drinking water and sewage disposal in the exploding squatter cities of Egypt or Brazil.

The inhabitants of the shadow world range from wealthy arms dealers to street peddlers, but what all who live or work in it have in common is the hiding of information—particularly financial information, but also information about the movement of products, the disposal of wastes, and the uses or abuses of their fellow humans and their planet.

Whether information is hidden deliberately and systematically or simply through the inattention of public authorities, the very nature of this shadow world makes it impossible to know with any accuracy just how much key information has become invisible or inaccessible. But some studies that have been done in particular countries or industries give an idea of the magnitude. In most countries, anywhere from 20 to 65 percent of all economic activity consists of black markets, organized crime, or unregistered work. The shadow economy subsumes a wide range of sectors—some of them opportunistic and dangerous, some of them made up of honest people who simply live below the radar of public scrutiny and beyond the reach of government protection and regulation. All told, a reasonable estimate would be that as much as one-third of the world's human activity is in the shadow. The result is an enormous *informational black hole:* a situation in which large amounts of critical data are disappearing into private sequestration and do not come back out.

Historically, people analyzing world affairs have tended to shrug off such holes—as, for example, in tables of global economic statistics which appeared in past decades with fine-print qualifiers like "except China," or in data on output that ignored all work done by women in households. If Texaco drills for oil, it's the energy industry at work, and it gets huge financial concessions from government; but if tens of millions of impoverished women spend hours each day gathering wood from an ever-shrinking global supply to provide fuel for cooking or heating, it's not counted as part of the energy industry—so there's neither support for the workers nor monitoring or regulation of the effects of their work on the planet. In 1996, the UN Development Programme estimated that the uncounted work of women—in food or energy production, education, or home-based piece work for large electronics or clothing manufacturers (employers that don't list these women on employee records and, thus, avoid paying employee benefits)—adds up to an astounding $11 *trillion* per year. In a world of expanding population and consumption, and its resulting pressure on the planet's carrying capacity, such data can no longer be overlooked. It's not just the lives and rights of exploited women or urban squatters that are at stake, but our collective ability to slow our destabilization of an already fast-changing environment—and to adapt successfully to those changes we can no longer avert.

Put it together, then. We say we live in an information age—yet paradoxically, our individual knowledge of what that information adds up to is becoming dangerously scattered and skewed. What are certain to be the most momentous events in the history of humanity are being marginalized; critical data concerning those events is being sequestered by specialists who either don't see the larger picture or don't know how it connects to the larger picture, or perhaps don't even care; we know less each year than we knew the year before; and large segments of human activity are disappearing from view altogether. Most public officials either don't know what's happening or don't want to talk about it. It is not an information age we have entered, so much as an age of *scattered* information—and shattered vision. To cope successfully with the immense changes we're now facing requires that we quickly find ways of seeing the whole—of re-grasping the world that is exploding from our reach.

3

THE SHOCKS OF SYNERGY

EVEN as the immensities of the megaphenomena begin to dawn on us, some of their greatest impacts will still take us completely by surprise. The mere fact that more and more devastating hurricanes, floods, fires, droughts, epidemics, and other predictable effects of the spikes have begun to send shock waves over the world will assure that *emergency* responses to those kinds of events are ramped up. But the real shocks could turn out to be effects that even the most vigilant of us fail to see coming. To make our vigilance more effective, we need to seek clues as to how such ambushes occur.

Throughout most of humanity's thousand-century experience, real surprises—truly astonishing events like the sudden appearance of an English sailing ship off the Aborigine coast of Australia—have probably been very rare. In the centuries before the spiking, because new developments came so gradually and infrequently, there would rarely be any big surprise in what happened, but only in when and where. Long before there were guns, for example, you might have known that if you loaded your horses and took a trade caravan over a certain mountain pass, you could be surprised by robbers. It was only the timing that made you nervous (will it happen this time?), not the thought that those robbers might do some kind of thing you'd never heard of—that, for example, while you prepared to defend yourself with your sword, they might slaughter you with a kind of weapon you had never seen. In the sixteenth century, when the Spanish Conquistadors shocked the Aztecs with horses and guns, such shocks were still rare. Only relatively recently, in the past one or two of those thousand centuries gone by, have truly unanticipated *kinds* of events occurred often enough to convince us that many more of them are now on their way.

It's also only relatively recently that people have sensed that anticipating events can do any good anyway—that we can intervene and change the way things turn out. Before the idea of probability was introduced by the French mathematician Blaise Pascal more than a century after the Aztec defeat, it would have been much harder to even *think* about reducing the risks of bad outcomes by taking precautionary measures. A certain fatalism—the gods will do as they will—prevailed. And even now, the idea of intervening and changing the future before it happens is a recent enough development not to have fully registered in popular perceptions. Look at the preoccupation with tabloid "prophecy"—the predictions of assassination or apocalypse that are presumed to have been written in the pyramids, or the Bible, or the Dead Sea Scrolls. These tabloids sell in far greater numbers than any scientific journal ever has. Groups that have vested interests in distracting us from the implications of the spikes have cynically exploited the fact that so many people seem not to have grasped that the future, in fact, is still very much in our own hands.

One of the most cynical exploitations occurred after the Stanford biologist Paul Ehrlich wrote in his 1968 book *The Population*

Bomb that "In the 1970s, the world will undergo famines—hundreds of millions of people are going to starve to death." In the 1980s, when that catastrophe turned out to have been mitigated (only tens of millions starved), groups that opposed policies to stabilize population attacked Ehrlich for being a "doomsayer"—despite the fact that it was warnings like his that helped spur the social and agricultural interventions that prevented that outcome from being realized. Ehrlich was thus marginalized from a central figure in biological science to a stigmatized extreme. But thousands of researchers now understand that Ehrlich's forecast conveyed an important warning, and that its purpose was not to prophesy but to provoke lifesaving action.

Yet, even with the growing confidence we now have in our ability to forecast, plan, and *change* probabilities, our perceptions tend to be impaired by the fragmentation of knowledge, and by the specialization that enables each of us to know more and more about less and less. You can see that impairment even in the discussions that strategic thinkers have with each other. In 1997, for example, the International Finance Corporation (IFC), a branch of the World Bank, circulated an internal document listing various kinds of surprise attacks that could disrupt international investments. The indicators included assassinations, strikes, riots, armed attacks, war casualties, political executions, and terrorism. These are, of course, the familiar surprises—today's versions of the sword-wielding robbers who might or might not attack on the well-trod mountain pass—that our established institutions continue to view as the main threats to global security. There was no mention, in this document, of such probable surprises as plagues of pesticide-resistant insects, massive hemorrhages of topsoil into the oceans, or important rivers going dry—or of how such events might interact. Yet those events, too, constitute forms of human-instigated violence, as destructive and as shocking in their final effects as any perpetrated by terrorists or rogue governments.

Within a year after that report was issued, the IFC was taken completely by surprise as the "Asian Tiger" economies of Thailand, Korea, Malaysia, and Indonesia were staggered by an avalanche of environmental and financial disasters. The World Bank had received repeated warnings by environmental economists, and even by some of its own analysts, that those investments were fundamentally unsound. But the bank's loan managers had been too caught up in the fervor of their established mission—running the world by moving around big infusions of money, as though it were manna from heaven—to pay much heed.[1] Within a few years, as commercial investors followed the World Bank's lead, there had been a rush to inject capital into countries not accustomed to assimilating it. When the erstwhile tigers suddenly looked more like a band of small cats that had been pumped full of steroids, the bank's other arm, the International Monetary Fund, was asked to lend a hand with reviving them.

It's no wonder that even the most conservative investors in the

1 ☙ In 1992, in the aftermath of the Earth Summit publicity, the World Bank accepted a recommendation by its industry and energy department that from then on, it would only make loans for projects that included measures to improve energy efficiency. Better efficiency, of course, meant reduced CO_2 emissions. Yet, two years later, an investigation by the Environmental Defense Fund (EDF) and the Natural Resources Defense Council (NRDC) found that of 46 power plant loans then being processed by the World Bank, only two met the bank's own requirement. Three years after that, in 1997, energy analyst Christopher Flavin reported that "fossil-fuel projects may be flowing through the Bank's pipeline faster than ever."

global economy seem to sense now that despite the end of the Cold War, the unseen dangers are not receding after all, but growing. Notice that more and more of the conservative affluent are moving into gated communities, hiring private police, and gutting government in favor of private enterprise that concentrates wealth.[2] While they share an ideological optimism about what their high-paying, high-tech industries can accomplish, they're uneasy about short-term security. They see no evidence that companies like Microsoft, Monsanto, or Mitsubishi have done anything to stem the rising tide of social unrest. For a short time in the mid-1990s, they may have thought the spread of the good life to the developing world would be a rapid and easy thing—as evidenced by their heavy investments in Asia. Investors upped the flow of private capital to developing countries from about $2 billion in 1970 to $244 billion in 1996, before the bubble burst. The knockdown of Asian economies was a sober reminder that stock-market gains can be a shell game—that to bet on the plans of a go-go Indonesian or Indian government, when the majority of their people are in extreme poverty, is to go out on a very long limb. When Asian economies make short-term recoveries and investors in the flush of relief go right back to their bullishness, the weight on the limb they've climbed out on only grows heavier. If real collapses come, they'll come all the more as a shock.

If there is any doubt that even with all the information and intelligence systems we have, there could be truly world-shaking surprises in store for us, consider the recent past—the period

2 ☙ By the beginning of the 1990s, US citizens were spending nearly twice as much on private security as on public police. The nation's total expenditures on police in 1990 came to $30 billion, whereas $52 billion was spent on private security.

during which the four spikes have begun their eruption and disruption:

- The *Internet* did not exist in 1980, and though it is dependent on telephones lines, was not anticipated even by AT&T.
- The *AIDS pandemic* was unknown in 1980, but within decades of its appearance had ravaged a whole generation of Africans and killed at least 17 million people on all six inhabited continents.
- The *fall of the Soviet Union*, without war, though dreamed of by Western governments for four decades, took the CIA and other intelligence agencies completely by surprise when it actually happened.
- The *resurgence of once-conquered diseases* took the World Health Organization by surprise.
- The *decline of the world nuclear industry* took electric utilities, governments, and the industry itself by surprise.

There have been, perhaps, other equally profound changes that don't make the same kinds of headlines, and of which we are not yet fully aware. These could include sea changes in human mental programming, or in capabilities to manipulate our own genetic future—not to mention their unpredictable side effects. But if history has taught us anything about large changes, it is that while we usually see in retrospect that there were warning signs, the signs are often wildly misinterpreted. Whatever the surprises turn out to be, the spikes will render our vulnerabilities to them immensely greater.

The most frequent causes of misinterpretation are the kinds of fragmented vision that have resulted from specialization, acceleration of change, and truncation of news. If you look at an incipient change and think only in linear terms about its direct implications—and don't look at how it alters the broader picture—you're very likely to be dead wrong. The nuclear engineers and investors of the 1970s, for example, saw a technology that could take tiny amounts of uranium and produce huge amounts of pollution-free electricity—and naturally, they expected to create an economic revolution. They did not anticipate the enormous complications, safety hazards, disposal problems, local resentments, and spreading fears and outrage that would dog them and eventually bring them to a halt. They did not expect the catastrophe at Chernobyl, or the unenterable dead zone it would leave in Ukraine for thousands of years to come. They certainly did not anticipate that the number of nuclear plants under construction in an energy-hungry world would fall from a peak of 30 per year in the mid-1970s to 2 per year in the 1990s.

Moreover, while some signs of impending trouble are shrugged off or misinterpreted, others are missed altogether—and the surprise is total. Yet, that's rarely because the underlying mechanisms are secret (Hiroshima bomb) or unknown (Ebola virus) or too carelessly monitored (Chernobyl accident). More often, we don't see signs for the same reasons of fragmented vision that cause us to misinterpret those we do see. For two decades, the Soviet Union prepared biological weapons that now appear to have been even more unthinkable than the nuclear ones, but we were too focused on the nuclear specter to register the biological one. Or, to put it differently, a *few* people saw the problem, but they weren't the ones who had the power to make intervening decisions—or to transmit the information effectively to people who did.

The metaproblem of the surprises we have experienced so far is that we can't see 360 degrees, or at least not all at once. We can't swivel our heads like birds. And even if we could, we'd be handicapped by the fact that to swivel 360 degrees still produces vision only in a circle—and the plane of the circle would have to be rotated through the three dimensions of a sphere to produce full visibility above and below as well as behind. And even then, you couldn't see inside your head; you couldn't see events of molecular or submolecular size; you couldn't see events outside the optic focal range, or over the horizon, or around corners, or at distances great enough for the intervening dust to scatter the light. Not all at once. But even if you could, would you necessarily know the meaning of what you see? A hawk, for all its acuity, may fail to apprehend the meaning of the gun barrel it sees aimed at it.

What we *can* do is assemble montages—the mental equivalents, say, of looking at a painting by the Los Angeles artist David Hockney, in which a series of shifting perspectives evokes a sense of moving through a complex landscape. If the montage is assembled in steps, we may be able to anticipate some of the synergies that could be driven by the combined force of the spikes in an already stressed biosphere.[3]

Anticipating, of course, does not mean prophesying or predicting. Once again, to emphasize that elusive distinction between predictability and *probability*, imagine you are caught in

3 ❧ The term "synergies" is used here to refer to the mutual exacerbation of two or more events, in ways that result in a vicious circle, or escalating loss of control. A generation ago, when I was growing up, the paradigm for such outcomes was the nuclear chain reaction. But a chain is a linear metaphor, apt for the age of the Industrial Revolution and its mechanistic twentieth-century aftermath. Now, in the new biological age, we know the synergies to be more like webs.

a tornado. A brick chimney is ripped from a neighbor's house and comes hurtling toward the spot where you are crouching. No one can predict what the outcome will be. A brick could strike your head—or you could escape untouched. What is certain is that, once that chimney begins its flight, the *risk* to you—the *probability* of disaster—suddenly rises in a way you have no time to measure but have no doubt is real. In each of the following scenarios, then, first look at the situation that now exists—and then consider the risks.

SCENARIO 1: OGALLALA AQUIFER, NORTH AMERICA

The settings for most stories about the American West have usually been the most visible features of the land: the mountains, mesas, canyons, and High Plains. The story told here is taking place out of sight, in a vast underground lake that stretches from South Dakota to Texas. The Ogallala Aquifer is a feature of the land that most of the early Caucasian settlers never even knew about. They knew they could dig wells, but probably never dreamed that the water they drew was coming from a body nearly a thousand miles long. It is, in fact, the largest body of fresh water on Earth.

The ground above the Ogallala is one of the richest tracts of farmland on Earth—the great American wheat belt. The United States is the world's most prolific producer of grain per capita, and it has the largest food surplus. A society that has a large surplus of food can afford to have more of its people working in jobs other than farming—jobs that may be important to developing new technologies, industries, and wealth.

The wheat belt is heavily irrigated with water from the Ogallala—pumped up and distributed over hundreds of square miles of those famous "golden waves of grain." In fact, according to

Sandra Postel of the Global Water Policy Project in Amherst, Massachusetts, the Ogallala supplies about 30 percent of all the ground water used for irrigation in the United States—or it *did*. As it happens, the Ogallala is not like a reservoir or a river, replenished regularly by rain. It is largely a fossil aquifer—the water left from the melted glaciers after the last Ice Age. It's been there for a long time, but when it's gone, it's gone. And, it is now being depleted rapidly; in recent years, the level has dropped and farmers have had to drill deeper to bring water up. In the United States as a whole, farmers and ranchers in the 1990s were drawing 20 billion gallons more water each day than is replenished by rainfall. In Northwest Texas, by the early 1990s, one-quarter of the Texas share of the Ogallala had been depleted. That was before the staggering drought that struck Texas in the summer of 1998, when temperatures climbed over 100 degrees Fahrenheit for 29 consecutive days.

When there's drought, as will likely happen more often now because of the carbon gas spike, the level drops faster because the soil dries faster and more irrigation is demanded. As water is depleted in particular parts of the aquifer, farmers over those parts are forced to cut back on irrigation. Some land is returned to dryland farming, which is less productive; other land is abandoned. This decline is already well underway. More than a third of the land irrigated in Texas in the 1970s had lost its water and gone to cracked ground by the 1990s. It's not just abandoned high-rises in Manila and abandoned warehouses in South Philadelphia that now haunt the industrialized world, but abandoned farms in Texas.

As we move into the future, and drought becomes more chronic and severe, these effects will likely spawn others. As farming is cut back, the native grassland is slow to return because over decades of monoculture farming and overgrazing by cattle, there has been a steady loss of the once-deep topsoil, not to mention the seeds of the native grasses that once thrived on it. The degraded land dries easily, and as the wind sweeps the plains it generates clouds of dust. Some of the oldest farmers may recall the Great Dust Bowl of the 1930s and wonder if it is returning.

Where farming continues, the greater heat will make crops more vulnerable to pests. One reason is that the warmer environment allows pests to expand their habitats northward, into areas that were previously just a little too temperate. A slight temperature increase can produce a large increase in range, and when the pests first migrate there are no natural defenses against them and they may do heavy damage. Heat stress also makes the plants more vulnerable to the pests, *whether or not* the pests have increased in number. All that causes further declines in crop yields. The farmers respond by applying heavier doses of pesticides, or by switching to more potent formulations. This proves to be an increasingly short-term and desperate expedient, as insects have become more and more resistant to pesticides over the years.

The pesticides do relatively little to stop pest damage, except temporarily. Moreover, the pesticides kill off many of the birds that are the predators of pests. In fact, by the 1990s, farm pests developed such high resistance to pesticides that they were consuming about the same percentages of the world's crops as they did in the Middle Ages. But while fighting a losing battle against insects and weeds, the pesticides have potent effects on *people*. (A single generation of Texans can't make selective adaptations to a toxic chemical, whereas during the same span of time a particular population of crop-eating insects might produce a hundred generations, enabling it to become completely resistant to the toxin.) In the next two minutes, as you read this, another 100 people will die of pesticide poisoning somewhere in the world. With a few more years of spraying, that toll could rise—adding to the blows already inflicted by bad weather, poor yields, and dust. The rising virulence of pests, coming atop to the lack of water, erosion of soil, and abandonment of land, cuts yields even more.

Meanwhile, the population spike is adding pressure from the demand side. The US population is creeping upward more slowly than that of many other countries, but it is still expanding by 3 million a year—equal to adding another city the size of Detroit, with all its needs for farmland, water, materials extraction, and waste disposal, each year. Moreover, as resource inequities have worsened, immigration is increasing. In the US sun belt, there are large influxes both of domestic migrants from the overcrowded north and of Mexicans and Central Americans from the heat-stricken south. As population growth pushes pavement east from Albuquerque and Denver, and west from Oklahoma City and Kansas City onto the High Plains, still more of the land is taken out of agriculture for housing tracts and shopping centers, as it has been taken out for suburban development over the past half-century. Whatever new development doesn't take farmland takes natural habitat, accelerating extinctions and further weakening the web of life.

In the region of the Ogallala, the bottom line—for the moment—is that while rising population has increased demand for food and water, the cascading effects of drought, plague, land cutbacks, and a declining enthusiasm for the farming way of life have all depressed food production. In this scenario, during the early years of the twenty-first century, the United States loses much, if not all, of its grain surplus. In doing so, it also loses much of its ability to exert geopolitical power—and security for its own people—through grain exports. But this is only the beginning, because it does not yet include the impacts of events elsewhere.

SCENARIO 2: NILE VALLEY, AFRICA

Something similar happens in North Africa, but with massively different proportions and impacts. Egypt's dependence on the Nile River is total. For millennia, with a small population and a large water supply, the country maintained a sustainable agricultural system by using the annual natural flooding to irrigate crops.

This system was so productive that for centuries, in addition to feeding its own people, Egypt served as a primary source of grain for the Roman Empire. When human population began to spike in the twentieth century, Egyptian planners decided to dam the river in order to control the flow and expand the irrigated area. In the 1960s, the Aswan High Dam was built about a thousand miles upstream (south) of the river's mouth, and for a while the plan seemed to work. Egypt's crop output expanded. But even as it did, two unexpected results became apparent. The first was that, paradoxically, even as the amount of food Egypt could produce expanded, its ability to remain self-sufficient—an ability it had maintained for 7,000 years—was lost. Within a few years after the Aswan's completion, the country that had supplied bread to Rome became a "net grain importer"—having to buy wheat from other countries. The situation recalled a warning Thomas Malthus had written more than a century and a half earlier: "Population, when unchecked, increases in geometrical ratio; subsistence only increases in arithmetical ratio." By doubling its food production capacity, Egypt had simply bought itself a little time in which to quadruple its population. The Nile, which on a map looks uncannily like an upside-down tree, had in effect been used like a tree in which the branches are required to bear more and more weight. The heavier branching off of water for irrigation allowed a larger population to depend on it. But that larger population was a larger weight being pushed farther out on a weakening limb, because it was unsustainable for reasons that had not been foreseen even by Malthus.

The second result of the damming, little noticed at first but increasingly apparent over the ensuing decades, was that it stopped the flow of silt, which for millennia had replenished the fields and built up the great Nile Delta, which is the heart of Egypt's food production. Geologically, the delta depends on silt buildup to live; like any living thing, it has to be fed.[4] The replenishment had already been somewhat diminished after the British erected some barriers in the river a century earlier. But the High Dam finished the job. With the silt-flow stopped, there was no buildup to compensate for the steady crumbling of the delta's banks into the Mediterranean Sea—an erosion that until then had seemed insignificant. By the 1990s, Egypt's richest land was disappearing into the sea at a catastrophic rate. Borg-el-Borellos, a village that once thrived on the coast of the delta, now lies under the sea 2 kilometers from shore.

With global warming, cropland dries faster and more of the irrigation water evaporates before it can reach the plants' roots. From its sources to the Mediterranean, the Nile flows through

4 ❧ Some scientists argue that in certain respects, and perhaps even fundamentally, ecosystems *are* living things, or macro-organisms. The movement of silt in a river is like the movement of nutrients in an artery. Yet, according to the International Rivers Network, when a large dam is built to trap a river's water, typically 98 percent of the river's silt is trapped as well.

some 4,000 miles of desert. The amount of water available downstream continuously declines. By the late 1990s, it had dropped so much that there was barely enough to irrigate the delta, and almost no water at all reached the mouth of the river to flow into the Mediterranean Sea.

As we move into the near future, the upstream countries will be forced to respond to the demands of their own rising populations, exacerbated by their own problems with increased drying. One branch of the river (the White Nile) flows all the way from Uganda in central Africa, the other (the Blue Nile) from the highlands of Ethiopia. In the early twenty-first century, all three of those countries will experience rampant population growth. UN projections in the late 1990s forecasted that Ethiopia's population will nearly triple between 1995 and the year 2030, from 56 million to 154 million. Even if the UN projection is unrealistic, Ethiopia is in deep trouble. Either the projected growth occurs and results in calamity, or it doesn't occur because the calamity has already materialized in the form of a massive starvation and population crash. Remember, Ethiopia is one of those countries where we first heard about famine—and now its demand has increased far beyond that which prevailed in the mid-1980s, when 300,000 of its people died of hunger. Yet, its ability to produce or buy food is now less than it was then.

The same situation will prevail in Sudan, where stability has been further eroded by the long civil war that has ravaged that country's ability to organize. The allocation agreements among the Nile countries will come under increasing strain, as the water becomes more precious. Sudan and Ethiopia will complete many of the more than 100 new dams that they had planned or started building during the 1990s. Less water than ever will reach Egypt, as more than ever is lost to upstream evaporation and withdrawals. The amount reaching the delta will fall to a trickle. No water will reach the Mediterranean, and salt will build up to crop-killing levels in the delta soil.

Egypt, with a huge impoverished population, will find itself imploding—facing a dire shortage of water and food, as well as of land on which people can live.

Of course, this problem is not entirely new. Egypt has been a major importer of grain since the 1970s. But now it will need to buy much more grain from the United States. Only, this surging need will be appearing just as the US capacity to export has dried up. It will be the end of the river for Cairo. But not the end of the story.

SCENARIO 3: SOUTHERN THAILAND

When the global economy began its boom in the early 1990s, Japanese and Western investors went after the newly industrialized countries (NICs) in a feeding frenzy. Bankok, Thailand was regarded as an especially promising investment environment—an ambitious, fast-growing city full of cars and motorcycles, international hotels, alluring consumer goods, and hustle. In 1991, the World Bank held its annual meeting in Bangkok to celebrate the Siamese Tiger's arrival as a major player in the global economy. The Thai government, for its part, demonstrated its commitment to modernization by evicting 200 families from their homes in order to widen the streets the delegates would be using during their brief stay.

If Bangkok epitomized the Asian Tiger economies at their most robust, it also epitomized the four spikes at their most rampageous. By the early 1990s, the cars clogging the city had brought on one of the world's worst cases of urban arteriosclerosis. About

300,000 more vehicles were being wedged in each year, most of them forced to creep at speeds of such poor energy efficiency (internal combustion engines have inherently poor efficiency to begin with) that they enveloped the city in a huge plume of toxic chemicals and carbon gas. Recent research shows that the lead in this plume may be causing losses of IQ in as many as 70,000 of the city's children.

Even so, the industrialized life was intoxicating. By the middle of the decade, as Bangkok spread out into the countryside, some Thais were apparently beginning to regard their ancient agricultural heritage with ambivalence, or at least with impatience. On one hand, the country was struggling to feed its population and needed all the rice it could produce. On the other hand, it was becoming clear to the Thai elite that they could get more support from the World Bank and private investors—and make more money faster—by using prime land for industrial development rather than for farming. Bangkok lies on the Chao Phraya River Delta, on which Thailand is heavily dependent for food, but as the city expands, more and more of the delta is being paved over. By 1997, 20 percent of the Chao Phraya watershed was developed.

The Chao Phraya River flows south from the central highlands where Thailand borders Burma and Laos, and passes through Bangkok on its way to the Gulf of Thailand. The farmers of the Chao Phraya basin are still poor—not yet a part of the brave new world of gleaming office buildings and cars. Some, however, have been offered a quick form of relief from their poverty. If there is a girl of 14 or 15 in the household, the family may be visited by a recruiter from the city, who explains that there are job opportunities for girls in the booming tourist industry. If the parents agree to let their daughter go to work in Bangkok for a year or two, she'll be able to make 25 times as much money as she can on the farm. If they give their consent now, there will be an immediate cash payment.

The offer is irresistible. The girl is taken to Bangkok and, much to her surprise, forced to work as a prostitute servicing the international sex-tourism trade. As recounted by environmental historian Aaron Sachs in *World Watch*, the trade first gained prominence during the Vietnam War, when the US military worked out an arrangement with the Thai government allowing US soldiers to come ashore for R&R (Rest and Relaxation), which gave the sex industry what amounted to official sanction. As the industry boomed, it was given tacit encouragement by the World Bank. In 1971, reported Sachs, World Bank president Robert McNamara, "without specifically mentioning the sex industry, urged Thailand to supplement its export activities with an all-out effort to attract rich foreigners to the country's various tourist facilities. After all, spending by US military personnel on R&R in Thailand had quadrupled between 1967 and 1970, from about $5 billion to about $20 billion." McNamara had to have been well aware of what kind of "tourist facilities" he was talking about, as he had been the US Secretary of Defense at the time the R&R agreement was made. In any case, by the mid-1970s, Thailand had 20,000 brothels and other sex-industry establishments. By 1994, about 800,000 Thai girls were being pressed into service. Government agencies, including police, turned a blind eye.

Meanwhile, Thai officials were under pressure to do much more to increase export revenue. One strategy was to modernize the country's fishing industry—to emulate what Japan had done with its fleet of high-tech ships that function not just as giant fishing boats, but as high-speed, mobile food-processing factories. In the 1990s, Thailand became a major producer of canned tunafish. Available in supermarkets everywhere, Thai tuna became one of the world's great protein bargains—it was selling in the United States in recent years for just over $2 per pound.

But this price, like the price of cheap gasoline or cheap lumber, did not reflect the full cost to the world. For several years, it had been apparent that global food production from land-based farms wasn't keeping up with population growth, in part because expanding cities like Bangkok were taking land away from farms. In almost every country, dependence on seafood was growing. By 1998, worldwide, people were eating twice as much fish as beef, and most of the fish was coming from the oceans. Sales of Thai tuna soared.

But fishing hadn't fully closed the gap between rising population and flattening grain production, because ocean fishing too had reached its limit. By 1997, according to the UN Food and Agriculture Organization, 11 of the world's 15 major oceanic fishing grounds had gone into serious decline as a result of overfishing. More ominously, as pickings got thinner the fishing ships were dropping their nets deeper and taking species once rejected as "trash" fish—species that a few years earlier would not have been worth the cost of processing them. Despite strong demand, the fishing ships were finding it more and more difficult to fill their holds.

As we move this scenario to the present, and on into the near future, a dangerous biological unraveling has begun. As the Thai fish factories become more desperate to meet their quotas—and more willing to haul up any fish they can, of whatever size—several things are happening simultaneously. The small species are normally the food for larger ones, so as the nets reach down the food chain to the smaller varieties, the larger ones lose their sustenance and begin to die off. At the same time, in their efforts to take more of the smaller fish, the floating factories also haul in more of the juveniles of the larger species—thus undermining future fish populations as well as exhausting the existing ones. That, in turn, not only steals food from future *human* generations to feed the present, but pushes more oceanic species to the brink—or over the brink—of extinction. The reflexive response of the fishing fleets, at least for a time, is to redouble their efforts to use ever more sophisticated technology to take even more of whatever living matter still survives.[5]

Of course, that practice, like the extraction of more and more water from the Nile, can't go on for long under any circumstances. But with global warming, what has been a worrisome decline can turn into a free-fall. In Bangkok, where we can anticipate that politicians will plead with marine scientists for explanations, the scientists may irritably point out that the problem is not a lack of answers, but of a willingness by decision makers to act on them in time. The scientists may recall, for example, that back in the late 1980s, some of their colleagues discovered that ultraviolet (UV) radiation coming through the ozone hole was killing large amounts of phytoplankton—the small oceanic organisms that are the primary photosynthesizers of the world and the *basis* of the oceanic food chain. While the ozone hole is a different phenomenon than global warming, it is an overlapping problem—caused by the same kind of unre-

5 ☙ This response is not unique to the fishing industry—it's also happening with forests (the average size of trees cut has gotten smaller as mature trees are depleted), farms (which have spread to more marginal and erosion-prone land as populations expand), and mining (in which lower-grade ores are extracted as reserves become depleted, thus increasing the volume of waste and pollution).

strained dumping of chemical wastes into the air. The news of this threat to the food chain had been disregarded off by the world's fishing nations, most of which had heavily subsidized their fleets in the race to extract whatever they could before it was gone. In 1997, further research had found that in addition to reducing phytoplankton growth by as much as 5 percent, the excessive UV radiation was damaging the development of crab and shrimp larvae, as well as of young fish. In 1998, still another major study had found that the whole web of oceanic life was beginning to unravel. That news had made barely a ripple in either the global media or the deliberations of policymakers. It, too, had been quietly marginalized.

Meanwhile, however, the Thais discovered another, more lucrative source of income: farming shrimp. All along the southern coast, during the 1990s, mangrove forests were cleared and factory ponds were built to raise shrimp for export. The industry was controlled by mafia-like companies that embraced capitalist free enterprise with an anarchic zeal much like the kind that was busy gutting the economy of Russia. In the village of Ao Goong, on Phuket Island, the native fishermen who had for generations made their livings by catching wild shrimp from the Gulf of Thailand were driven out of business by the factory ponds (the managers of which also conveniently dumped their huge outputs of waste on the villagers' land, killing their palm trees and poisoning their drinking water). Wild shrimp normally spawn in the shallow water surrounding mangrove trees, but with the mangroves bulldozed for the farmed-shrimp operations, the wild shrimp are dying off. The villagers, like the people whose houses were razed for the World Bank meeting, have been sacrificed—their local economic well-being subordinated to the larger goal of increasing their country's national export income.

At the same time, shrimp farming is also helping to further decimate ocean fish populations, because ocean fish are hauled up in bulk to make fish meal for the shrimp. Each year, 5 million tons of ocean fish are ground up, worldwide, to make enough feed to raise 1 million tons of farmed fish and shrimp. Ecologically, this practice is as absurd as that of the overweight American who uses a power mower on his lawn then pays money to push a rowing machine. Again, the market costs don't reflect the real costs—to the ocean and to people. And those costs have to be paid, sooner or later.

The Thai economy came crashing down in 1997. It would be a mistake to conclude that the "rescue" loans from the IMF will make any real difference in the long run. They may temporarily raise investor confidence and keep the con game going. But the IMF still operates by the discredited doctrine that a country can grow its way out of poverty even when that growth is achieved mainly by a liquidation of natural capital. It's a doctrine based on the same false accounting that props up the consumption boom worldwide. And so the Thais continue to generate cash by selling off, at liquidation prices, what is most valuable to them—their inland forests, mangrove forests, oceanic ecosystems, and even some of their young women.

Back to Texas, near future. Producing less grain also means producing a lot less beef, since a pound of beef takes seven pounds of grain to produce. In the North Texas region of the Ogallala, cattle ranching has been a primary industry, but now this too is in trouble—not only is the feed for grain-fattening of livestock getting prohibitively expensive, but the grasses for grazing are drying up. Prices rise for meat, as well as for fish. And because water shortages are being experienced elsewhere too, food prices are rising all over the world. US producers target distribution where the market will bear these prices. The politics of privatization, along with the declining role of government, assure that very little will go to "food aid" via US or UN subsidies—no matter how many babies with distended bodies now appear on TV news. About the only food exported from the United States goes to markets for the wealthy in other countries. There are wealthy elites in Nigeria, Indonesia, India, and Mexico who will pay whatever the market will bear. The Texas cattle ranchers stay in business by selling less meat to far fewer people at far higher prices. The world's poor—including US poor—get less and less.

Egypt, near future. The country is caught in a tightening vise between swelling population and declining food supplies. Its population is rising by 1 million a year. But its supply of food has been cut by the mutually exacerbating disruptions of reduced flow of water from upstream, heightened evaporative loss, and shrinking cropland. Economic refugeees spill across the Gaza into Israel, and across the Mediterranean into Europe. In countries like Italy and Germany, already stretched by heavy immigration from Turkey and eastern Europe, strained hospitality turns to rising hostility. Some European politicians disingenuously demand restoration of national borders. Some wealthy Europeans, Israelis, and Egyptians emigrate to the United States and take high-tech US jobs—made easier by the fact that a few years earlier, in the 1990s, US high school students had ranked near the bottom of the world in math and science tests.

United States, near future. Hostility to refugees increases here too, as combinations of poor immigrants from Mexico and more affluent ones from Europe, Asia, Africa, and South America drive the influx over the panic threshold for those who once assumed the country was theirs. The increasingly strident demands of brown faces in local white-majority communities becomes muddled, in public perceptions, with the demands of brown faces abroad (seen on nightly TV coverage of spreading food riots in Cairo or Addis Ababa) for a more equitable sharing of the US wheat and corn once so abundantly shared. Food distribution becomes a political issue strongly controlled by global food-and-drug companies, in much the same way that for half a century US energy issues were controlled by fossil fuel interests. The issue, however, now isn't just about exports, but about domestic food distribution. That, in turn, opens a political Pandora's Box.

For decades, any hostility between rich and poor in the United States has remained remarkably subdued, considering the scale of the inequities. Maybe the idea of social mobility has defused it; the poor don't want to rock the boat of the rich if they believe they may soon board it. And popular mythology in the United States is replete with stories confirming that possibility. As long as the poor or the sinking middle class aren't actually aching with hunger, they can daydream. The dreams are sustained by a range of magical devices: state lotteries, commercial sweepstakes, casino gambling, giant jury awards, and the gargantuan salaries of sports and entertainment stars. All of these suggest that it is not uncommon to jump from making $10,000 a year to making $10 million. Or sometimes, $10 million in a single day or minute. In the 1990s, millions of poor black kids *identified* with megawealthy sports or entertainment stars. Why would they want to shoot those stars down, when they might soon hit the jackpot themselves?

With the evaporation of the US food surplus, however, the means of maintaining that myth of quick mobility could shrivel. At present, few of the working-class poor are inclined to point fingers and complain that an AT&T executive has a million-dollar salary only because he helped eliminate the jobs of a thousand of *them,* and thus increased the stock values of all those rich folks out in the surrounding suburbs. But political demagoguery—or even just ordinary political opportunism—could change that. In the 1990s, many American blacks were angered by a news story linking drug trafficking in their neighborhoods to CIA activity in Central America. That allegation was later widely reported to have been proven false, leaving many skeptics uncertain what to believe—and effectively blunting their anger. But it would be quite another matter to deny that the rich have prospered at the expense of the poor. "Even though the GNP of the United States grew considerably during the 1980s, three-fourths of the gain in pretax income went to the richest one percent," noted economist Paul Hawken. That trend continued through the 1990s, not only in the United States but worldwide. It won't have been the failed promises of "trickle-down" policies that finally impel the hands of the deprived to reach out and *take,* but the bitter understanding that you can't be another Michael Jordan if you can't eat well enough to stop the pain in your belly. One of the main reasons the mobility dream worked is that people still got to satisfy their hunger for cheap sugar, fat, and meat. When the cheap food runs out, so does the narcotic of instant appeasement.

The omnipresence of cheap calories in the industrialized world thus works much like the availability of cheap gas for cars. Subsidizing the energy of the body is, in the end, no different than subsidizing the energy of the technologies that extend the functions of the body. At both levels, the quick gratification buys off the consumer and pushes away any suspicions that underneath, a starvation is taking place. An affluent society can joyride in its cars even as its ecosystems are starving, and an impoverished population can get fat on empty calories even as it is suffering abnormally high rates of heart disease and malnutrition. In American cities, where the average child drinks more liters of soft drinks in a year than of water, the heavy promotion of cheap sugar and fat recalls the situation that prevailed in ancient Rome, where surplus grain was distributed free to help keep a discontented populace quiescent. The eventual disappearance of the surplus, as a result of depletion in outlying provinces—including Egypt—weakened the empire's grip and became a factor in its decline. As will be seen, this has been a recurring pattern throughout human history. The road that led to Rome now leads to Texas.

All this is greatly oversimplified, and the actual events will be messier. Other interactions between the spikes—resulting in other destructive synergies—are inevitable, although we can't know for certain what they will be. But scientists have sketched out some of the more far-reaching possibilities. Many are now concerned that the spreading use of genetically engineered plants in farming could result in the appearance of "super-pests"—weeds or insects that can no longer be controlled and quickly destroy much of the human food supply. A related worry is that as more and more of the planet's bird, reptile, and amphibian populations are wiped out, natural controls on insect populations will collapse. A third concern is that as human interference with the Earth's atmospheric chemistry continues to intensify, climate change could become much more sudden—even more sudden than the spikes themselves. Wallace Broecker, a scientist with the Columbia University's Lamont-Doherty observatory, recently warned, "The climate system is an angry beast and we are poking it with sticks." And science writer William K. Stevens noted in the *New York Times*, in January 1998, that "a growing accumulation of geological evidence is making it ever clearer that in the past, the climate has undergone drastic changes in temperature and rainfall patterns in the space of a human lifetime." The six months following that statement proved to be the hottest six months—by a larger single jump than ever before—in the recorded history of humanity. What we will see in the coming years will likely include not only scenarios like those sketched out for the Ogallala, the Nile, and Thailand, but others we can only dimly envision. In short, as the primary impacts of rising population and consumption fan out, the complexity of the ripple effects grows exponentially and explosively. As noted earlier, the impacts don't progress in straight lines.[6] Instead, the lines branch and intersect, becoming vastly more numerous as they progress.

Complexity also escalates as the effects leap rapidly from the realms of the physical and biological to the economic and social and then to the psychological—and ultimately to the human spirit. As they do, the science available for tracking them becomes murkier. As a result, many of the analysts who have been studying the physical impacts of climate change have stopped short of discussing what that will mean to people. A few social scientists and economists have ventured to outline the effects in broad terms, some of which are summarized in the scenarios above: there will be more people uprooted from where they live, more backlash against refugees, more rear-guard attempts to close borders, and harsher questioning of rich-poor inequities. Few have attempted to study—not to mention finding the institutional mandate and funding to study—what happens when,

6 ❧ The world was perceived much more simply in the Cold War era, when there was one global enemy and the prevailing metaphor of the threat it posed was that of a row of dominoes falling. Many of today's kids, with their hyperlinked Network access and multidimensional video games, may not even know what a domino is.

as is likely, these physical and social sea changes also bring a rising psychological tsunami—of frustration, resentment, anger, angst, depression. What we can be sure of is that some of these waves will intersect, with often unpredictable results.

While the outcomes are impossible to predict in their specifics, scattered data offer suggestions of what we face. For example, media coverage of events like the Red River flood, focusing on heroic responses and the will to rebuild, turn our eyes from the evidence that over time, massive misfortune drags people down. A study released in 1998, for example, found that victims of natural disasters have much higher suicide rates—highest of all for victims of floods. Suicide is the tip of the iceberg of social destruction wrought by depression, and it is becoming a plausible hypothesis that spreading incidence of disasters could cause a whole society to become abnormally depressed. Already, by now, five of the ten leading causes of disability worldwide are psychiatric.

In the two countries where population is highest, the synergies of the spikes have already precipitated an epidemic of individual suicides. In India, which is projected to have the world's largest national population by 2010, and where the government has gone all out to increase GDP by increasing exports, the GDP has grown feverishly—and yet poverty has grown in lockstep with it.[7] One of the big exports is cotton, which provides the T-shirts and bed sheets for the burgeoning populations of other countries. To ramp up exports, Indian farmers were compelled to re-

7 ❧ "In absolute numbers of people trapped in poverty, India is still the world's poorest nation," observes Worldwatch Institute researcher Payal Sampat. UN Development Programme data indicate that "More than 500 million Indians earn less than $1 a day (many of them less than five or ten cents a day) in purchasing power."

place their traditional diversified agriculture with monoculture cotton. The natural pest controls provided by a complex agriculture were replaced by heavy applications of pesticides. But the pests, as noted earlier, are growing resistant to pesticides as fast as the chemical companies can formulate them.

The farmers depended on getting big yields to pay the loans they had to take out to finance this new, high-tech kind of farming, but the yields were decimated by caterpillars. In early 1998, as reported by the *Wall Street Journal*, a farmer named S. Sailam, who lived with his pregnant wife and two children on a cotton farm in the southern Indian state of Andhra Pradesh, took the pesticide that was no longer working in his field and squirted it down his throat. In the first two months of the year, more than 100 farmers in Andhra Pradesh died in this manner.

In China, where the majority of farmers are women, a similar epidemic of despair is under way. In the village of Shizhuqiao in Hunan province, reporter Lijia MacLeod recently visited the home of a man named Luo Fangrong, whose wife Xiahua—driven to despair by the same kind of failing battle with pests—swallowed a bottleful of pesticide and collapsed, leaving him and their three daughters. "Women are dying in the countryside in alarmingly high numbers, and the pressure of China's fast-paced economic development is a major culprit," reported MacLeod in the *Far Eastern Economic Review*. Data compiled by the World Bank, World Health Organization, and a Harvard University research team show that while 21 percent of the world's female population lives in China, 57 percent of the world's female suicides are occurring there. Most of them are farmers, betrayed by an overwhelming compounding of stresses.

Of course, the disabilities inflicted by a failing environment or economy are not always manifested by self-immolation or immobilization. A person hit by too many stresses at once may search for blame, and in a tangled web that search can follow many threads. Some will blame the rich and join populist groups; some will blame government and further dismantle the governments that are really their principal remaining protection; some more analytical people will (perhaps correctly) associate the breakdown of systems with the overweening confidence we had in technology, and will (mistakenly) assume the answer is to destroy technology. One of the nastier surprises of the twenty-first century may be the emergence of more people like Ted Kaczynski, the Harvard-educated "unabomber" who espressed his rage against technology by, ironically, sending bombs through the mail, and who was generally treated by media as an anomalous loner.

It's not unlikely, however, that in many cases the rage of the frustrated will be unfocused—or focused on the nearest visible target, like the attack of the Texas tower sniper in the 1970s. Along with the unabombers and Texas snipers, we could see more religiously outraged World Trade Center bombers. And, as high-powered weapons technology percolates into the hands of private groups and individuals of all ilks, such attacks could escalate into generalized assaults on worldly society, as in the Tokyo subway gassing.

Meanwhile, in a world where mass communications are not only far more powerful than in past eras but far harder for governments to control, one of the most ascendant occupations on the planet could be that of the charismatic demagogue—the politician who knows how to add fuel and air to anger. As poverty, resource shortages, and crowding all worsen, so does anger. In Egypt, instead of lone Kaczynskis, we could see a surge of anticapitalist fury aimed at the global corporations that have exploited the rich-poor divide and that have declined to let American grain come to the slums of Cairo. Or aimed at those who have tried to keep impoverished Egyptians out of the suburbs of Tel Aviv, Brussels, or Berlin. The demagogues in Egypt could harness Islamic unrest in a way that makes the Bin Laden terrorist campaign seem marginal.

The demagoguery in countries like Egypt, Pakistan, Afghanistan, and Iraq could be echoed in previously stable countries like the United States, where apprehensions about rising anti-free market fervor, combined with unsettling levels of seemingly random attacks from within an increasingly fragmented society, would play into the hands of still more opportunists—including the familiar kind known as Republicans and Democrats, whose platforms would bristle with increasingly draconian anticrime, anti-immigration, antiterrorist measures. These, in turn, would further distract the political process from focusing on root causes—the human expansion and its impacts on the planet's climate and life—and keep it focusing policies on maniacs.

But again, these high-profile outlashes are only the tip of the ice storm. The biggest problem, now, is that under the surface, there's a latent anger or denial in hundreds of millions of us. As twentieth-century humans, inheritors of the Renaissance and the Industrial Revolution, we felt the world was ours. Now we feel betrayed by our own destiny. Many of us feel betrayed by technology. We feel let down by God.

In the foregoing scenarios, it becomes harder to envision the details as we follow the threads further into the future—and as we proceed from hard physical and biological phenomena to observable social phenomena to speculation. In the actual event, many of the details will be different and maybe even impossible to imagine from our present perspective. Remember, with each year, the distance traveled in magnitude of change gets longer—and so the perceived speed at which we move gets faster. By 2010, we will be moving at a speed so great that it will stagger us. The important thing now is not to try to predict final outcomes, but to identify the forces driving them. The stories are still being written, but the forces are real.

There will be a strong desire, in many, to say this outlook cannot be right. Those who put their faith in "technological optimism," especially—including the hired guns of industries that have made a gospel of growth in general, and of the growth of the fossil fuel industries in particular—will predictably argue that human ingenuity will solve any problem that arises.

It's easy, having had our fingers burned as many times as we have, to underestimate the capabilities of human ingenuity to finally make things right. But to sanguinely point to those capabilities as a reason to say we don't need to worry or to make revolutionary changes in our behavior in order to save ourselves, is to risk making two potentially calamitous mistakes.

The first mistake is to assume that our uses of technology so far constitute a history of steady advances for civilization. In fact, many inventions that were celebrated when they first appeared have set us back more than they have advanced us, in ways that came as complete surprises—nuclear power, asbestos insulation, breast implants, ozone-depleting refrigerants, and DDT, among them.

The second mistake, in arguing that there's no need for "drastic" action now, is to assume that if we use our ingenuity to correct for the few really large setbacks, such as that of the Nile Delta drying up, we'll be OK. As that argument goes, the basic economic system is sound, and it would be foolish to unnecessarily rock the boat. But the drying of the lower Nile is not an exceptional case, or even the worst of its kind. The Yellow River Delta in China, which supports an even larger population than Egypt's, is drying up too. So are the Amu Darya Delta and the Colorado River Delta. Deltas are drying up all over the globe. Similarly, the Ogallala Aquifer isn't the only fossil aquifer running out of water; so are the ones that supply water to Saudi Arabia and Libya. More than 100 countries are now undergoing some degree of desertification, and at least 22 of them now don't have enough fresh water to meet the needs of their present populations.

The impulse to think we don't yet need to undertake an unprecedented human mobilization is not just one of ignorance about what's happening now, but about what we think happened in the past. The logic of the techno-optimists is that human ingenuity will solve all problems in the future because it always has in the past. The talk-radio and newspaper-column pundits appear to really believe that, partly because they live in places that have not yet been overwhelmed by floods, squatter populations, mafias, food shortages, electric grid failures, or epidemics. But history, which leaves us now with the legacy of a human population too large for Earth to support without liquidating its resources, does not support the view that our ingenuity is supreme.

4

AMBUSHES OF THE PAST AND WHAT THEY TELL US

THERE have been few cases in which a human society was threatened by forces other than military invasion and was sufficiently aware of the threat to mobilize. Until the advent of the four spikes, such threats were only sporadic and regional—disturbances that only with hindsight were recognized as signs of a coming disaster. On the other hand, whole civilizations have disappeared because they apparently did not know, until too late, that the way they lived was unsustainable.

Most of us grew up reading textbooks or hearing accounts that made only brief mention of these disappearances and offered only the vaguest explanations of them. Some of them we never read about at all. Attention was focused on the civilizations that prevailed—the ones that became our heritage. The others, with a few charismatic exceptions like the pre-Columbian Maya, were forgotten. Our consciousness of history is one of civilizations competing for dominance in an ongoing drama for which the geographical settings are an unmoving field of battle, like a chessboard.

This view is reflected in the *Histomap of World History*, a chart sometimes seen hanging on classroom walls, which depicts the expansions and contractions of civilizations beginning at the top, four millennia ago. It's a useful device for showing that civilizations, like individual organisms, seem to have periods of robust growth, decline, and death. It also suggests, perhaps, that there may be a parallel between the tendency of young humans to not really grasp that they can die (as seen in the excessive risk-taking of adolescent drivers, for example), and a young civilization to be oblivious to its own mortality. But the *Histomap* is also highly deceptive in one respect: it is designed so that at any given moment, some civilization—whether it be the Egyptians, the Assyrians, the Persians, the Greeks, or the Romans—is always robust or dominant. The implication is that the Earth is a fixed, unchanging arena, like that chessboard—always there to be occupied by whichever human player will claim it. Our conventional view of history does not suggest that a time might come when the chessboard rises up and takes back the space and no society wins.

The techniques of historic reconstruction have vastly improved, however, since the times when the human story was written only by the winners and only on the evidence of what the winners remembered or inferred from the records of antecedent winners. We have added the ability to analyze what people ate, what illnesses or injuries they suffered, what they made their clothes from, what kind of environment they lived in, and what changes took place in that environment. Through analysis of DNA, bone damage, tooth wear, pollen, the shapes of tools or weapons, the contents of ancient cooking fires, and the silica skeletons of plants, we can do a virtual cultural autopsy. We can reconstruct—or at least sketch out—the histories of peoples who left cryptic monuments or records and of those who disappeared altogether. What we find is that the waning of civilizations is not, as implied by the *Histomap*, simply a matter of weaker societies being shoved out by more powerful ones. In some cases, at least, civilizations were killed—or were weakened internally—by faults that had little or nothing to do with the powers of competing human societies.

Such reconstruction has identified a surprising number of cases in which a society vanished not because of conquest but because of some kind of environmental disaster—often self-inflicted. In other cases, conquest was only the last act—the main reason for the fall being the environmental self-annihilation that preceded it.

- The civilization of ancient Sumer thrived for more than two millennia, during which eight of the world's first cities arose in the region of present-day Iraq. The city of Uruk had a population of about 50,000 by the third millennium BC. Expanding population was supported by a system of intensive irrigation by water channeled from the Tigris and Euphrates Rivers. But the hot climate caused heavy evaporation of the irrigation water, leaving a gradual buildup of salt in the soil. Wheat, which was the main crop, can't grow in salty soil. By 1700 BC, the once-rich land was failing and the civilization had collapsed. Crop yields had fallen by 65 percent. The Sumerians could no longer feed an army and fell to invasion. All eight cities returned to dust.
- The Tehuacán Valley society arose 7 thousand years ago in the south-central region of what is present-day Mexico. This was a neolithic society—one of the first to make the transition from hunter-gatherers to farmers. The shift took millennia, with the productivity of the land gradually becoming more intensive. The Tehuacános began with cultivating maize (the forerunner of corn), then added horticulture and hydrohorticulture—the cultivation of

vegetables and fruits, both in soil and in water. Around 3000 BC they begn using irrigation. By 2000 BC their crops were about 50-percent irrigation-dependent, and by 1000 BC they were about 80-percent irrigation-dependent. As history has proven again and again, heavy dependence on irrigation is risky. If the climate is hot, salt accumulation can ruin the soil as it did in Sumeria; if drainage is poor the land can become water-logged. If there are no horses or cows to produce manure (there were none of either on the North American continent at that time), the soil may eventually become exhausted. Exactly what happened to the Tehuacános' sophisticated agricultural society is not certain. But a few centuries later, their crop yields declined and their civilization came to an end.

- The city of Teotihuacán, in the Valley of Mexico (centuries after Tehuacán) rose with the development of a bigger cob of maize between 400 and 300 BC. It may seem surprising that something as seemingly trivial as the size of a corncob could form the basis of a civilization. But consider, as noted earlier, that what technology does is to extend the powers of the human body. If today's technologically-intensive society gets its energy from fossil fuels, then a human-energy-intensive society like the Teotihuacán would get its energy from firewood and food. If food energy can be concentrated—as happens when the size of a corn harvest doubles or triples for the same amount of work—then the society has surplus energy. It can feed a diverse work force—and an army. But if the energy is being produced in a way that is unsustainable, the system will break down. Something in the Teotihuacán system proved ultimately unsustainable—whether it was the cycle of deforestation and soil erosion that would later bring down the Mayans or the salinization that had undone the Sumerians. At its peak in the sixth century, Teotihuacán was the capital of the largest empire to exist in the Americas before the arrival of the Spanish. It is believed to have had a population of 150,000. It had lasted about a thousand years, but after 600 it went into decline and, having been weakened from within, was easily conquered from without. In about 650, Teotihuacán was sacked and burned by the Toltec, just as Rome had been sacked by the Vandals about two hundred years earlier. The city was overgrown by jungle and forgotten.

- The Roman Empire maintained its power by using outlying provinces to supply food in much the same way that modern industrial states use the developing world to supply oil, aluminum, and timber. The city of Leptis Magna, for example, was the center of a productive agricultural area in what is now Libya. It became part of Rome in the second century AD and was the site of numerous architectural wonders, including a 12-mile-long aqueduct. As the empire grew, however, the land was overcultivated, and production gradually fell—in some areas by 50 percent. Much of the land eventually turned to desert, and Leptis Magna was abandoned. In other regions, the empire's demands for expanded food production pushed cultivation into ecologically vulnerable hills. As the hillsides were cleared of trees they became heavily eroded. The city of Antioch, the Roman capital of Syria, now lies under 28 feet of eroded silt.
- The Harappan, or Indus, civilization arose around 2300

BC along the Indus River in what is now Pakistan. Like the ancient Egyptians, the Harappans utilized the annual flooding of their river as a means of irrigating a very arid landscape. Cultivating wheat, rice, dates, melons, peas, and cotton, they supported the cities of Mohenjo-Daro and Harappa, which were the center of an empire more extensive than present-day Pakistan. Around 1750, the Indus empire came to an end. There are no known written records of why, but one theory is that as the rising population put increasing pressure on the farmers to produce food from a fixed amount of land (the land that could be reached by the flooding), the nutrients in the soil were depleted. Another is that the Indus changed its course and left the cities high and dry—though that would not explain why the civilization died instead of moving. A third theory is that the Indus communities were wiped out by a great flood. A possible clue is the fact that the bricks of which Mohenjo-Daro was built were fire-hardened—meaning that large amounts of timber must have been cut to fire the clay. Deforestation is known to greatly worsen the severity of river flooding. Much of the original city of Harappa, like Antioch, is still buried under silt.

- The Mayan civilization, in parts of what are now southern Mexico and Central America, lasted over three millennia. By 200 BC the city of Tikal had arisen in what is now Guatemala; by 600 AD its population had reached somewhere between 30,000 and 50,000. Over time, the Mayans cut down much of their forest—for firewood, construction, and the manufacture of lime plaster for ceremonial buildings, as well as to clear land for agriculture. The deforestation caused severe erosion of topsoil, leading to failures of the crops needed to support such a populous society. After 800 AD, the Mayans went into an abrupt decline. Within a few decades, their civilization too was gone—their cities covered by jungle.

- The civilization of Easter Island evidently began with a very small migrant group and grew over a number of centuries to a peak population of 7,000 to 9,000 in the mid-sixteenth century, around the time of the first European exploration of the Earth. This civilization lasted a thousand years, but the population eventually outgrew the island's natural carrying capacity, which—like that of the Earth as a whole—is finite. Once that happened, the civilization went into decline and collapse.

On a global scale, there was probably no threat to humanity *as a whole* until less than a century ago. The first inkling, perhaps, came in 1917—not with World War I, but with the outbreak of a particularly virulent strain of influenza, which killed more people during the course of the war (about 30 million) than the combat and bombing did. War, at that time, probably could not have threatened the whole human future. For war to threaten a territory required a massive mobilization of equipment, and many parts of the world remained untouched. For the flu, all it took was for a single person to arrive on foot, coughing.[1]

A second global threat came a quarter-century later, with the development of a capability for intercontinental nuclear war. A

1 ❧ There were plagues in earlier centuries, of course, but the natural barriers of geography effectively prevented global transmission; the Black Death of the fourteenth century did not sweep Asia or the New World, as it probably would have if there had been extensive air travel then.

third was the thinning of the ozone layer and the resulting increase in UV radiation. A fourth was the capability for biological warfare—first called "germ" warfare, but later renamed to include viruses and other forms of biological destruction.

These four have occurred in just the last 1 percent of the 80 centuries or so that civilization has existed, so there's been relatively little time to absorb their importance into our cultural attitudes, or even to fully assess what that importance is. The 1917 flu was given little attention by historians and was largely forgotten over the following eight decades. Only in the 1990s, with the rise of global concern about disease vectors, was it rediscovered. The nuclear arms threat galvanized the world for decades, but then it too was largely forgotten with the end of the Superpower arms race. But it may quietly have grown rather than receded, as France, Iraq, Pakistan, and India have taken turns showing off their weapons. And the nuclear threat, like the long-forgotten flu of 1917, has only mutated into a different form—a part of the underworld economy that could pose a more insidious danger than the old one. The biowarfare threat, too, was prematurely forgotten, and only in the last few years have officials awakened to its implications. But there too, there's reason to believe public concerns could fade—not because the threat will have receded, but because the warnings may.

The ozone-hole threat[2] was the last of the four to emerge and still remains in public consciousness—yet there are signs that it too is on its way to being prematurely dismissed. The Montreal

2 ❧ The threat comes from the accumulation, over half a century, of chlorofluorocarbon chemicals (CFCs) from millions of aerosol sprays, refrigerators, and air conditioners into the upper atmosphere, where they destroyed large patches of the ozone layer that shields us from UV radiation.

Accord on Ozone Depleting Substances, unlike the Kyoto climate treaty, was a relative success. It didn't call for an eventual reduction of emissions by a paltry 5 percent, like the Kyoto fiasco; it banned the main shield-destroying substances outright. But as with carbon dioxide, it's easy to forget that banning new emissions of a dangerous gas isn't at all the same as clearing out the half-century's accumulation of gases already hanging over our heads. Even as news stories in the late 1990s announced that CFC emissions were dropping, the accumulations in the atmosphere were still rising. They also failed to mention that while legal CFC sales were ending, black-market sales were booming. Experts projected that it would take until about 2050 to restore the shield to its normal condition—by which time, the UV may have triggered irreversible unravelings of some of the planet's most vital systems. The suppression of phytoplankton, the base of the ocean food chain, continues unabated. In short, all four of these global threats remain unresolved, even as we're beginning to stop worrying. Three of the four—the nuclear, biowarfare, and CFCs—feed industries that are becoming entrenched in the global shadow economy. The fourth—the flu—remains a scientific enigma; 80 years after it struck, scientists are still unable to explain why it struck down young, healthy adults even more savagely than it hit the weak and elderly.

As the first century of globe-wide vulnerability came down to its final year, it was clear that in the thinking of most policymakers, the degree of risk posed by global threats *other* than military ones had not yet sunk in. Even the nuclear specter, which had been an obsession as a military threat, had never attracted much interest as a threat to global health and ecology. Trillions of dollars had been spent on "national security" aspects, but little on the larger questions of what happens when radiation rides the wind and water currents, enters the carbon cycle, and invades food—considerations that make it irrelevant whether you are American, Russian, or Indian, or indeed whether the nuclear release is from an exploded bomb or a power plant meltdown.[3]

The most important observation to be made about these twentieth-century global threats is that they are all, in one way or another, manifestations of the four spikes. The flu recalled the Black Death of the Middle Ages, but was also a precursor of the spate of new and resurgent diseases of the recent past and present, which in turn are driven *by* the spikes. The nuclear threat was the century's most horrific example of how technology can release Pandora-like dangers every bit as impressive as the services they provide—the specter of thermonuclear war being an illustration-in-a-flash of what the carbon and extinction spikes do just a bit more slowly. (If this seems like hyperbole, consider that another 200,000 square kilometers of the world's land, an area about the size of Florida, is turning to desert each year.) The biowarfare threat is nothing less than the risk of a deliberately inflicted escalation of the extinctions spike—bioextinctions in a bottle.

Through a miracle as worthy of our reflection as any other in human experience, no global threat has yet done its worst: we still live. There are important things to be learned from these threats, particularly regarding the way we are inclined to ignore them—and the way they mutate and regroup. But because their stories

3 ❧ Those of us who grew up under the nuclear threat were acutely aware that even if we weren't killed by the direct hits, we'd likely die of the fallout—the radioactive poisoning of our air, water, and food—wherever we lived. But that awareness had little effect on government policy or spending priorities, which focused obsessively on the military face-off.

are not yet finished, we can only speculate about final outcomes. For more definitive understanding of what can happen as such ignoring of threats persists over time, we're thrown back to the examples of regional loss, in which the stories did end. What clues can be gleaned, as to just how we have let things run too far out of control, and what we can do to regain stability in time?

A scan of the circumstances in which civilizations declined or fell, looking past the military/political conflicts and into the ecological underpinnings that give or take away the vitality of every civilization, offers a few potentially vital clues.

The remains of vanished civilizations are a little like the belongings left in a house whose residents have departed suddenly and left no forwarding address—a place mostly empty but with scattered enigmatic remains suggesting haste.

Haste is not the same as speed, though if events continue to accelerate as they have in recent decades, that growing speed will inevitably induce more and more hasty responses. Haste is what happens when the time horizon for planning retracts to the point that people either no longer see the full cycles of which their present actions are segments, or no longer feel able to take them into consideration.

A telling example can be found in what happened to the people of Rapa Nui, or Easter Island, whose disappearance posed one of the world's most famous mysteries. Rapa Nui is the easternmost island of the eight-time-zone-wide region of the Pacific known as Polynesia. When Dutch explorers under the command of Jacob Roggeveen arrived in 1722, they found a barren land on which giant stone figures—torsos of ancient gods—stood facing the sea. There were over 600 of these statues, and it was soon deduced that the stones had been moved long distances from quarries—but how? This was a century before the beginning of the Industrial Revolution. Later research would find that the average statue was 13 feet high and weighed over 12 tons. One, found lying supine, was 71 feet long. It was conceivable that such massive stones could have been moved on wooden rollers, but that would require great timbers, and the island was barren of trees. There were also no horses, or other draft animals. There were a few natives still alive on the island, though within a few minutes of Roggeveen's arrival their numbers had been reduced even further. (The natives who gathered to watch the Europeans come ashore were unarmed, but the Europeans marked the occasion by shooting into the group and killing 10 or 12 of them.) The few who remained had no skills in carving or transporting stone monuments and were reportedly so desperate for food that they had become cannibals. They could tell the Europeans nothing about the lost civilization they had presumably descended from: they were what might today be described as a post-apocalyptic remnant.

The mystery continued into the mid-twentieth century, when scientists using pollen analysis found that Easter Island had once been covered with large palm trees—relatives of the present-day *Jubaea chilensis*, which grows to heights of some 65 feet in Chile. Further forensics put the story together. The island had been settled by a group of 30 to 100 Polynesians, who arrived in one or two large ocean-going canoes around the sixth century. They brought chickens with them, along with sweet potatoes, sugar cane, bananas, taro, and yams, for planting. They also, evidently, brought a few rats that had been hiding in their cargo. The colonized rats, like other bioinvasions taking place the world over, would play a significant role in the new colony's eventual fate.

To clear land for planting, the settlers used the same slash-and-burn approach that has been practiced by settlers for millennia, and that is now being used to clear rainforests in the Amazon and Indonesia. Trees were cut to build houses, to fashion fishing canoes, and to provide fuel for cooking. As the population grew, more forest was cleared for the larger crop yields needed, and the biggest trees were cut to make canoes that could go far enough out to sea to catch porpoises and tuna—which would have required going at least a kilometer out. And, more trees were cut to make flexible tracks for moving the stones for the statues, which were central to the people's religious and cultural life.

The population was divided into numerous clans, and erecting ever-greater statues seems to have evolved into a form of competition. But, like the building of ever-more impressive sports stadiums by rival cities in our own time—or, perhaps, the building of ever-more intimidating military arsenals by rival countries—this use of public resources proved unsustainable.

As the population spiked, and the pace of tree-cutting accelerated, the island was completely deforested. This set in motion an ecological unraveling that would later begin to be replicated on a global scale. As the largest trees disappeared, new sea-going boats could no longer be built. Carbon dating of the bones left from food remains shows that deep-ocean fish and porpoises disappeared abruptly from the diet around 1500—right around the time the population peaked and began to fall. But by then the people could not wait for the smaller trees to grow large enough to build new canoes, because those trees were now needed for cooking fuel.

When the last existing boats finally fell apart, we can surmise, the fishing ended—all except for what could be done from the shore. When the houses could not be rebuilt either, the people moved to caves. Civilization went into reverse. The deforestation caused the soil to erode, and the exposed ground—deprived of both shade and moisture-holding topsoil—reverted to a kind of desert. Food production fell sharply. Researchers believe that in earlier centuries, the people of Easter Island had stored palm nuts as "famine food," for times of drought, but now the palms were gone. The rats, whose population had spiked along with the people's, now openly competed with people for what little food could be scavenged. The now unsupportable human population, deprived of both its sustenance and its aspirations (the statue-building had ceased, of course), collapsed.

The fatal moments, though probably not recognized at the time, would have been the felling of the last large trees for firewood. Those trees, put to a different use, might have allowed—had plans been made early enough—for a recovery of some kind. Suppose, for example, that the last generation of palm nuts had been used to begin a program of reforestation, instead of squirreled away and then eaten. At worst, the very last tree might have made an ocean-going canoe for a last journey for help from the distant Australs or Marquesas islands. But by then, we can infer, the future envisioned by whoever approached that tree extended no further than the meager meals that could be cooked with its wood, never mind any arduous trip undertaken in a last hope of saving the loved ones left behind. It may have been a rational decision under the circumstances, but it was also a fatal one.

Today, the same kinds of rational decisions are being made again, but on a vastly larger scale. Thus, China's leaders may judge it a rational decision for their country to build 500 new coal-powered electric plants now even though that will release disastrous amounts of carbon dioxide into the sky, because those leaders believe they can't wait for a long technological journey to develop alternatives. Or, Thailand's aquaculture mafias may consider it a rational decision to scoop five tons of low-value fish from the ocean to use as feed for each one ton of high-profit shrimp or salmon they raise, because their accounting doesn't have to include the net loss of four tons of life from the ocean for each ton they sell.

Many of the societies that have disappeared have left signs of such hasty departures from sustainable practices. The Indus, Teotihuacán, Mayan, Ethiopian, and Easter Island collapses were all triggered by deforestation, which led to soil erosion and crop failure. The Sumerian and later Mesopotamian and Teotihuacán collapses were preceded by profligate use of irrigation, which led to salination or water-logging of the soil and, again, crop failure and famine. When famine sets in, planning terminates: in the winters of 1316 and 1317, hungry peasants all over Europe took the last of their grain—the seed they would otherwise have used to plant the next spring's crops—and ate it.

The more recent falls, despite all the resources now available to stop them, have followed the same patterns. In every case, people driven to keep up with their growing population and consumption have undertaken whatever practices they thought would boost immediate output. They may or may not have known this was putting them in an impossible bind. Either way, these are the acts people are driven to make when they are running out of time.

A second set of clues is more speculative, but could be critical to our understanding of what happens to information that may be vital to a society's survival but may not be believed—or welcomed—by the people in charge. Did doomed civilizations have people who knew?

Documented cases of direct warnings are hard to find, perhaps because such warnings are threatening to the people in power, and the message (not to mention the messenger) may have been quietly snuffed. In some cases, warnings that might have been dangerous for a political adviser to make could be made in the form of religious or philosophical observations that might be less threatening to rulers. "None shall cut off a stream of water," wrote Philo Judaeus of Alexandria, the ancient city on the Nile Delta, during the first century AD. His advice seems to have been heeded for nearly two millennia, until the High Dam was built at Aswan.

But look more broadly at what happens to visionaries, and it's apparent that many have been either suppressed or disregarded. Plato, for example, warned in his *Critias* that deforestation and soil erosion were ravaging the land, and that "what now remains compared with what then existed is like the skeleton of a sick man and only the bare framework of the land being left." He apparently understood well enough the role of healthy forest in retaining fresh water, as he described the soil of an earlier time, "It was enriched by the yearly rains . . . which were not lost to it, as now, by flowing from the bare land into the sea; but the soil it had was deep, and therein received the water, storing it up in the retentive loaming soil."

That simple fact—that the roots and ground cover in natural forests are critical to the sustenance of soil, and therefore to the strength of the society that depends on that soil, has been disregarded again and again in the millennia since, with fateful consequences. Twenty-five centuries after Plato made his observation, a Cornell University agricultural ecologist, David Pimentel, turned it into hard numbers. Assembling data from all over the world, Pimentel found that naturally forested lands lose between 3 and 40 pounds of soil to erosion per acre, each year, while the natural soil-building processes of the forest accumulate more than that. But when forest is removed for timber, or grazing, or growing crops, the loss of soil rockets to between 7 and 16 *tons* a year—roughly a thousand times as much. On the steep hillsides of Ethiopia, Rwanda, Mexico, Tibet, or Peru, to which some of the world's poorest farmers have been forced to retreat, the soil is hemorrhaging at 40 tons per acre per year.

Warnings like Plato's can be recalled now because they came down to us from cultures that were exceptionally literate. It's likely that many others have been destroyed. And if that is so, it's likely that a general pattern of behavior among threatened human societies is to become more blindered, rather than more focused on the crisis, as they fall. Those who try to raise alarms may not only be stigmatized as extremists, but punished.

In societies where public opinion or sentiment plays a large role in public policy, as it did in ancient Greece and Rome, reactions to news that the regime is on an unsustainable course could go even further—to include attempts to manipulate opinion in a way that turns people from worrying about the future to submerging themselves in immediate gratifications. In short, if the policies of leaders encourage or subsidize the marketing of distractions or denial, they are often signs of trouble of a magnitude the controlling regime can no longer afford to air fully—or which, if it can't entirely hide, it will at least try to push to the margins of consciousness.

A third set of clues suggests that information gaps in key areas of governance have been as instrumental in bringing down civilizations as holes in the hull of a ship might be in causing a sinking. These gaps can be matters of inadequate scientific knowledge to sustain the technology being used, or they may be matters of key information being hidden or lost.

Inadequate science was a major factor in the decline of ancient Mesopotamia, one of the places where civilization first emerged from pre-agricultural ways. As hunter-gatherer populations gradually transformed themselves into herder-farmer societies, they became more able to stay in one place, build permanent structures and towns, and use surplus farm yields to sustain the builders, artisans, organizers, and others needed to run an organized society larger than a single family. The size of the crop yields determined the capacity of these first towns to support nonfarming occupations and to develop into complex, interdependent societies. For the first time in human existence, the family was not an independent self-sufficient economic unit—it was part of a trading system. As the towns grew, the demand for larger harvests increased, and irrigation—the primary means of expanding yields in arid areas—was developed. What the farmers of that time apparently did not understand—or understand well enough to convince their leaders to take remedial action—is the process by which irrigation in warm climates causes a buildup of salt in the soil. When their crops eventually failed, we can surmise that either they didn't know why or they left no record of why. And so, when the subsequent civilization of the Sumerian cities arose, it was doomed to repeat the same mistake.

At other times, the regime may have had the information needed for control quite in hand, then lost it. And, one way key information is lost is that it gets out of reach—the regime gets overextended and finds itself unable to keep up all the communications needed to maintain a stable system. Rome, in its wane, lost contact with too much of what was happening on its vast periphery. One of the things happening was that its grain-growing regions in Syria and Libya, like those of Mesopotamia millennia earlier, were becoming degraded. The Romes of today, as will be seen, have lost touch with much of what is happening under their own noses.

A fourth set of clues concerns the shifts of sovereignty that sweep human civilization from time to time. When the Roman Empire was broken by internal division and invasion and gave way to a feudal system, the geography of Europe was fragmented. The focus of historians and their students is on the political dissolution of the empire into rival kingdoms and fiefdoms, and what happened in the no-man's-lands between those jurisdictions is of little interest. To individual lords, it may have been of no interest—had, indeed, they known—that the continent was undergoing a massive deforestation and biological simplification.

The legacy of that fragmentation has become one of the modern world's great illusions—the view, widely held by people across the political spectrum from north to south and left to right, that Europe is a model of what civilization can be when culture and environment reach a mature state. We can close our eyes, envision the towers of Chartres Cathedral above a flowering countryside; Monet's painting of the boating party; the English gardens; the Tuscany of olive trees and vineyards. Even environmentalists are likely to think of Europe in pleasant contrast to the ravaging poverty and near-anarchy of West Africa, or the demographic enormities of India and China. Environmentalists may point to the stabilized population of Europe, the greenness of the French countryside, the high standards of European living. But what we all overlook, in such sanguine views, is that Europe only got to be the way it is—and has kept itself there—first by conquering and colonizing most of the rest of the world over a three-century period, and then by shifting from a colonial domination to an economic one. It is an arrangement by which the industrialized world continues to extract resources from the "developing" world (a great euphemism, since more people in that world are impoverished now than in that Imperial era), and to maintain its relatively pleasant condition via the wealth of its former colonies—now "trading partners."[4] And in overlooking all that, we also overlook why European explorers were so driven to colonize the rest of the world in the first place.

One reason was that the formation of medieval states had greatly accelerated the internal exploitation of Europe's own resources. The natural Mediterranean forest of beech, oak, cedar, and pine had been extensively cut for firewood and for timbers for houses and ships, as well as to clear land for agriculture. Many animals were hunted to extinction, and the countryside lost much of its biological diversity. Beaver, which were common throughout Europe in the early Middle Ages, disappeared from the continent, as did wolves and aurochs—the wild ancestors of cattle. Bison, which were common in Belgium and Germany, also disappeared. Cranes, auks, and other once-common birds were seen no more.

Today, the biological nakedness of Europe can be measured by the extent to which its major watersheds have been stripped of their original forest. In the Guadalquivir basin (Spain), 96 percent of the native forest is gone; in the Seine basin (France), 93 per-

4 ❧ In *How Much Is Enough? The Consumer Society and the Future of the Earth*, Alan Thein-Durning writes that in The Netherlands, "millions of pigs and cows are fattened on palm-kernel cake from deforested lands in Malaysia, cassava from deforested regions of Thailand, and soybeans from pesticide-doused expanses of Brazil in order to provide European consumers with their high-fat diet of meat and milk."

cent is. The loss has come to 89 percent in the Loire (France), 85 percent in the Ebro (France and Spain), 85 in the Garonne (France), 81 in the Po (Switzerland, France, and Italy), and 71 percent in the Rhine (Belgium, Germany, and France). Overall, according to historian Clive Ponting, no more than about ten percent of the original forests that once stretched from Morocco to Afghanistan even as late as 2000 BC still exist. If Africa, South America, and Southeast Asia were to undergo the same degree of biological simplification Europe has—and then to depend as much on the resources of other continents to sustain them—the consequences would be catastrophic.

Europe, in fact, followed some of the same patterns Easter Island did. The difference is that although large parts of what are now Lebanon, Libya, Syria, Sicily, and Spain were deforested and in some places turned barren, Europe still had enough trees left to built the boats—so to speak—to sail out for more goods. Those boats were the ones used by Columbus, Cortez, and the other adventurers who triggered what would become, in the sixteenth and seventeenth centuries, a massive inflow of slaves, furs, timber, spices, fruits, gold, and other wealth removed from other continents.

History's preoccupation with military and political restructuring might have caused us to overlook the ecological transformation of Europe in any case, but if that transformation had gone as far under the old Roman Empire as it was to go in the ensuing centuries, it would almost certainly have been more noticed; the Roman administrators might have expressed greater concern about the shrinking inventories of wheat, wood, furs, and other resources. During tumultuous changes of power such as occurred during the empire's decline, things fall between the cracks. Sometimes those things turn out to be more important than the things being fought over.

That's especially worth considering now, because the changes in power taking place now are both faster and more complicated than any that took place in the past. More things are falling between the cracks. Nations have assumed a vast array of responsibilities for managing various mutual problems, yet sovereignty is moving to other institutions in the meantime, and many of the nation-forged solutions could be lost in the shuffle.

What this means, for strategies dealing with global threats, is that it could be a mistake to depend too heavily on the mechanisms of national governments to provide the needed solutions. In the United States, for example, there are coalitions of environmentalists who have been feverishly focused on the battle to reform Clean Air and Clean Water laws. But in the overall scheme of things, the actions of the United States government over the next decade or two are likely to become smaller and smaller parts of the whole—not just because the United States is becoming a diminishing fraction of world population, wealth, and power, but because *all* national governments are losing some of the autonomy they once had. Witness the struggles of once-feared Russia to control its own business moguls, mafias, and thieves; the rapidity with which once-impregnable China has seen its intellectuals begin to communicate with the rest of the world via an uncensored Internet; and the growing constraints on even the most proudly autonomous nations by such supranational entities as the World Trade Organization or European Union.

As we agonize over the question of how to cope, we'd do well not to reflexively assume national government legislation is the final goal, but rather to ask: Where will the real power be wielded a decade or two decades from now? To answer that, it's essential to get a better grasp of how information is used to wield power.

Even the US intelligence agencies, the most heavily funded information-gathering organizations on the planet (but often the last to know), have now openly acknowledged that the management of information must now be regarded as a potential "weapon of mass destruction." The ongoing shifts in how we learn, or how we come to think what we think we know, could turn out to be among the most treacherous surprises of all.

5

DO YOU KNOW WHERE YOU GOT YOUR INFORMATION?

In September 1997, the *New York Times Magazine* published an article with this eyebrow-raising title:

The Population Explosion is Over.

At the time it appeared, human population was rising, worldwide, by the equivalent of another New York City each month—and it still is. No one could be more aware than the editors of the *New York Times* of what an enormous cost, both economic and ecological, is involved in supplying a population the size of New York's with enough uncontaminated water, uninfected food, unleaking shelter, uncorrupted police, uncontested sites for garbage disposal, unbroken sewer pipes, and adequate schools, among other things; in fact, their city has struggled with those tasks even when supported by one of the world's richest urban tax bases. Since the implication of the article's title is that whatever growth rate now prevails is not a problem, it would be interesting to know how the *Times Magazine* editors imagined, when they approved it, that the world is going to provide for an additional population that large, from scratch, with every new waxing of the moon.

"Provide," of course, doesn't just mean providing the land and building the infrastructure for a city of 7 million. It also means finding all the land needed to grow the food, filter the water, produce the lumber, crack the petroleum, manufacture all the goods that population consumes, and dump or recycle all the resulting garbage.[1] And even if the world's governments could generously manage to donate the resources for all that once, could they do it again (with a whole new set of resources) the next month, and the next, and the next? That's what they're faced with now.

If a week passed between the time the *Times Magazine* received the article and the time it went to press, the world population grew—births minus deaths—by another borough of Queens, or maybe even a Brooklyn, in just that time. Perhaps the editors who accepted this article weren't thinking about anything

1 ☙ A population the size of New York's may produce a Brobdingnagian pile of municipal waste, but the waste generated by the far-flung industries that support that population's share of the economy is far greater. For every ton of municipal waste produced by a city in the industrialized world, more than four tons of industrial waste are produced and disposed of in other places, to keep that city going—much of it in places where it makes its way into our fresh water, soil, air, and food chain. So, for every additional million people, the added demand on the planet includes not only the land needed to grow enough food for a million, but the land taken up by whatever mining, drilling, pumping, and dumping is needed to produce the buildings, streets, vehicles, refrigerators, furniture, clothes, and everything else used by that many people.

beyond the terrific stir it would produce. However, its publication very likely had another, perhaps unintended, effect. At the time, a growing number of educated people were starting to worry, perhaps for the first time, that the troubles afflicting Calcutta or Cairo—the economic and social strains of serious overpopulation—might be starting to spread into American suburbs. For almost anyone experiencing such worry, this article would have provided reassurance that everything is going to be OK.

The *Times Magazine* story is a troubling illustration of how the fragmentation of information subverts our ability to see the whole picture. Once, in a simpler time, we called such fragmentation taking information "out of context," but that phrase doesn't begin to convey the magnitude of the distortion a story now can produce without telling a single actionable lie. The article's author was Ben Wattenberg, a well-known writer on demographic trends. His argument was sophisticated, disarming, and—if you kept your focus narrowly on his numbers—enough to persuade a concerned reader that all the warnings about population have somehow been a big mistake. The gist of it was that the world's birthrates have been declining, and that therefore what is now a population spike will thankfully peak in another half-century or so and then decline. A few of his points are worth a closer look, because they illustrate how smoothly such distortion can be inserted into public subconsciousness.

Wattenberg correctly noted that experts cannot agree on what the likely population growth curve will be: he pointed out that UN projections plot three curves for the year 2050: a high (worst-case) projection of 10.7 billion people, a medium projection of 9.4 billion (later revised to 8.9 billion due to the impact of AIDS in Africa), and a low variant of 7.7 billion (later revised to 7.3 billion).[2] Because birthrates are falling in many countries, he suggested, a figure that splits the difference between the medium and low projections—or 8.5 billion people—is "what is most likely to happen."[3] At the time the article was published, the global population was still closing in on 6 billion.

There is no point in disputing Wattenberg about whether the population will pass 10 billion and keep climbing or peak at 8.5 billion and begin declining. What's troubling, though, is his apparent conclusion that if the math works out as he thinks it will, the problem is over. It's as if the pilot of a plane that had lost all its power were telling the passengers, "Don't worry, it may be scary up here, but it's safe on the ground—and that's where we're headed right now!" To say that projections show the explosion will be over half a century from now is to ignore what we have to go through to get from here to there.

What's highly seductive about neat mathematical projections—whether they be those of the World Bank, FAO, or the American Enterprise Institute, for which Wattenberg works—is the way they can turn half a century of human tragedy into a curve on a graph, and if the curve ends up as we hope it will, then what

2 ❧ The UN's 1998 revision gives new emphasis to the essential challenge of the population spike: that either we find ways to bring it down deliberately and humanely, or it will come down of its own accord—whether through famine, disease, sociopathic behavior, or ecocatastrophe.

3 ❧ The explanation for falling birthrates (where they are falling) is a widely accepted theory called the "demographic transition," which says that as women become better educated and more affluent, they have more career or lifestyle options—and are more likely to delay having children, or to have fewer of them. The UN projections of eventual decline are based on the assumption that levels of income and education will continue to rise everywhere.

has happened in between doesn't really matter. If the population *will* stop growing just half a century from now, then we can relax. The trouble is, that's just what we did half a century ago; when the trauma of World War II was beginning to recede, we relaxed. Everyone who could afford to do so began living the good life—having lots of kids, moving to the suburbs, putting some steaks on the grill, buying station wagons and TVs. It was during that half-century that the spikes of consumption, carbon-gas, and biological extinction shot out of control.

During the fourth decade of that half-century, when the global population had just passed 5 billion, we were given the *World Scientists' Warning to Humanity*—a warning that we had gone too far, that we were taking unconscionable risks. During the *next* half-century, in which even the optimistic Ben Wattenberg agrees population will climb by another 2.5 billion or so (the total population on Earth at the time of World War II), what will happen? And, for that matter, what about the half-century after *that*, in which even if population is declining as Wattenberg and others say it will, it could still take decades to get back to where it is now? In short, the *Times Magazine* article—its title notwithstanding—acknowledges that population over the whole next century, even under the most optimistic of projections, will be far larger than it is now, when things have already become dangerously strained.

While the broad brush of Wattenberg's picture is deceptive, the details are probably even more dangerous, because they create an impression of careful analysis. The article noted that in 1950 the average fertility rate (the number of children a woman had in her life) was five, but by 1997 it had dropped to less than three. It did *not* note, however, that when you expand from a global population of 2.5 billion people in which women have five babies each, to a population of 6 billion in which women have three babies each, the total number of babies *increases*. And as long as it does, the base population (and the number of women of child-bearing age) gets still bigger, and the momentum of the growth becomes even harder to stop. Until the fertility rate falls world-wide to about 2.1 (two kids to replace two parents), the base *continues to expand*.

The article also noted, prominently, that some countries have achieved population stability or even have slightly declining population. It mentioned Italy. But it didn't point out that those countries constitute only a small fraction of the world. Taking Italy's low birthrates as a sign that world population is stabilized is like taking the abundance of pasta in Italy as a sign that the problem of hunger has ended. In population, it would take 40 Italys to equal one India, and in India population is still growing fast. (In fact, the UN projects that India's population will increase by 572 million in the next 50 years—an increase equal to twice the total population of the United States, added to a subcontinent that is already near the end of its rope.)

In the *Times* article's view of population, like the World Bank and FAO view of global food production, there appears to be an unquestioning acceptance that the baseline conditions (the "ifs" of an if-then projection) will remain the same long into the future. Ironically, Wattenberg was sharply (and rightly) critical of the United Nations for doing this kind of mindless extrapolation in its population projections. Yet, in arguing that the UN is too pessimistic, he cited a list of factors pulling down fertility rates that *he* assumed will remain in effect. They won't. Because of the synergies between the spikes—the surprises that we'll try to anticipate but probably can't—many of his fertility-reducing "factors" will likely be irrelevant before another decade passes, say nothing of 50 years.

Wattenberg noted, for example, that as infant mortality rates decline, parents become more confident that their own children will survive. "When parents know their children will survive, fertility rates plummet," he wrote. So far, so good. But assuming that fertility rates will continue falling in a world where new diseases are appearing at unprecedented rates and old diseases are resurgent, and where water pollution is getting worse instead of better, is like assuming food productivity will continue to rise in that same world. If half a century of human overpopulation has caused our life-support systems to begin unraveling, then another half-century of even larger overpopulation offers no promise that infant mortality rates won't start rising again, reversing one of the key factors that enabled fertility rates to begin falling.

What made this article conspicuous wasn't that it conveyed a dangerous deception (we're all too used to that now), but that it did so with real data and reasoned discussion. We're accustomed to seeing most issues treated mainly as competitions, in which our interest is presumed to be focused on who wins and who loses. So, anything that goes substantially further—looking at numbers and analyzing factors—is presumed to be informed opinion. We get no warning that the analysis may be a fallacious concoction.

Seemingly in-depth pieces like this have a special function in our system of info-distribution: they sow the seeds from which all the empty soundbites and inanities of the rest of the media spring. Only a very small minority of the population reads upscale publications like the *New York Times Magazine*. But those who do are the ones who—like pollinators who know not what they carry on their feet—then spread key bits of the message into the rest of the media. They include, I would guess, many of the journalists who write for thousands of small-town newspapers, the people who call opinions in to radio talk shows, the TV producers, the civics teachers, and the county supervisors and school boards. Together, they are the people who set the unstated limits of what's reasonable and what's extreme. And, because they know they don't know subjects like population very well, they don't attempt to actually discuss the subject—they just know how to keep it in its proper place in a conversation. They go away from an article like this one with a *sense* of the subject—a sense that, for example, anyone who says population is a looming problem for the world is a "doomsayer" or an extremist "neo-Malthusian" who is motivated by "an agenda."

Once the seeds of an idea like this have been planted, there is of course no accountability for the results. People making quips on radio, or expressing opinions in newspaper columns, or adding parenthetical background observations to news stories, do not footnote the origins of their views, because the views are conveyed with no real content. These comments, in turn, get repeated in casual conversations. By the time what's left of the original message has spread out through the populace, it has been laundered—it is, for all practical purposes, untraceable and uncheckable.

If we do want to pursue where our general impressions originated, the key is to find the primary seed-sowers, note what core ideas they convey, and then determine who has a really strong interest in sowing those ideas. In the case of "The Population Explosion is Over," for example, the American Enterprise Institute espouses a doctrine that has come increasingly to dominate the beliefs of global leaders: that it is competition, more than cooperation or sharing, or any other paradigm of human interaction, that will bring the greatest reward. That is the organization's premise.[4] It's an idea, however, that fails to recognize the deep flaws in a system of enterprises that does not have to account for all its costs—just as the competing clans of Easter Island did not have to pay for the trees they cut or the soil they allowed to erode. The AEI gets its funding and its promotion from those who profit from this system. It has an ideological interest in not letting our enthusiasm for expanding global markets be dampened by worries about unsustainable population and consumption.

Look more closely at each of the spikes transforming our planet, and a pattern becomes apparent. For each of the spikes, the public reaction is dreamlike and slow—a vaguely troubled face or an obliviously unconcerned one or even a completely blank one. Among those who are troubled, there is confusion and lack of resolve. And not far away, a powerful and purposeful organization is working very hard to keep it that way. It may be a large corporate phalanx like the Global Climate Coalition, or a small ideological think tank like the AEC. But inevitably it's an organization with a vested interest, financial and psychological, in denying that the spikes (which they do not recognize as such) are urgent problems. As the problems loom larger, and harder

4 ❧ The AEI's mission, as described by the World Directory of Think Tanks, is "to preserve and to strengthen the foundations of a free society—limited government, competitive private enterprise . . . and a vigilant defense." During the Reagan administration in the United States, a White House official told the *Atlantic* that the AEI had helped get Reagan elected by making conservative ideas "intellectually respectable." The ideas it helped popularize include two, in particular, that have become driving factors in the environmental unraveling of the Earth: the "trickle-down" theory that letting the rich get richer through competitive enterprise would help lift up the poor, and the doctrine that growth is the engine of human progress.

to deny, the spokesmen for these interests shift smoothly to other tactics—such as dismissively acknowledging that there may be a problem, but assuring us that to deal with it effectively will require no major changes in the ways we do business.

The sowing of messages designed to make public eyes glaze over are not random. They show up wherever the status quo seems particularly threatened—as it was, for example, by the activist movements that led to the UN population summit, the global biodiviersity convention, and the global climate treaty. Often, these UN-led movements collapse under their own bureaucratic weight. But if they show signs of actually convincing policymakers to make major leaps and to enact new principles of human governance that will profoundly change how we work and live, the mission of the seed-sowers is to plant just enough confusion or indecision to make those policymakers pull back from the brink.

At the climate conference, for example, the GCC strategists evidently worried that despite their ad blitz in the previous weeks, last-minute emotional arguments about our children's future would "get to" the delegates and impel them to draft a treaty with real teeth. At any rate, on the next-to-last day of the negotiations, an article appeared in the *Wall Street Journal* with this title:

Science Has Spoken: Global Warming Is a Myth.

It consisted of an opinion by two Oregon chemists that the world's climate scientists were wrong. Like the *New York Times Magazine* article on population, it turned out to contain no substantiation for its curious title; it was completely out of the blue. But it was, doubtless, enough to cause confusion in Kyoto, as GCC lobbyists ran to the conference rooms waving faxed copies of the story.

The final effect of such stories, whether or not substantiated, can be to make decisionmakers wonder whether they really have enough information on which to act after all. The result, as intended, is hesitation. It is a hesitation that has stagnated global policies for years, while the oil continues to flow, the slash-and-burn clearing of forests continues to wipe out species, and a billion women of the Third World continued to be denied options. The tragic irony of this hesitation is that it turns the precautionary principle on its head. The instinct of conservatives, of course, is not to rock the boat too hard if we're not sure what we're doing. Yet on closer examination, the "boat" being rocked is not the industries we've created, but Earth on which they and we are dependent. The articles that call for "caution" thus confuse us not only about the science, but about where the burden of proof really lies.

Go back to that image of the plane that has lost its power. The "be cautious" stories being planted in opinion-setting publications are like the warnings of a frightened air traffic controller telling the pilot, "We're not really sure that any action you can take will bring in the plane, so you'd better play it safe and *do nothing*." And indeed, in the industries resisting rescue, there does seem to be an attitude that future historians will regard as having been, in the broad view, suicidal. More realistically, it is just extremely myopic. As long as the profits flow, even up to the final seconds, disinformation is put out to fend off interference. And so, as the spikes have penetrated more deeply in the past few years, in addition to "The Population Explosion is Over" (*New York Times*) and "Global Warming is a Myth" *(Wall Street Journal*), we have seen the following (among hundreds) appear: "Environmentalism and Sorcery" (Institute of Public Affairs, Australia —suggesting that environmental alarms are taken seriously only by the kind of credulous person who believes in sorcery); "Global Warming: Apocalypse or Hot Air" (Institute of Economic Affairs, UK—suggesting that warnings about global warming have no more basis than the prophesies of apocalypse appearing in tabloid newspapers); "Recycling is Garbage" (*New York Times Magazine*—suggesting that recycling is nothing more than a foolish fad, and that the most economical thing to do with garbage is to dump it); "Grow or Die" (*Your Company* magazine—cover headline of a special issue suggesting that the only way a business can survive is to keep expanding).

Who writes these things? One answer is that a whole industry does—a corporate public relations industry disguised as a news industry. The newsletter *PR Watch* describes a business of disinformation that began operating soon after Rachel Carson published *Silent Spring*, in 1962. Monsanto Corporation, with its heavy investments in many of the chemicals her book discussed, issued a book of its own, ridiculing her conclusions. Subsequently, more sophisticated campaigns were launched by other industries that were encountering problems with their public images. When medical researchers issued their first reports that asbestos could cause cancer, the asbestos industry launched a countercampaign that in retrospect bears a striking resemblance to the one mounted by fossil fuel companies of the GCC. In similar fashion, campaigns were launched to dispel public worries about dioxins, clear-cutting of forests, and tobacco.

Over the decades, the corporate PR industry found that it could be most successful by closely mimicking the news industry, so that its "press releases" and reprints of medical or scientific "journal" articles were indistinguishable from the real things. Publishers and producers found it cheaper to simply take what came over the transom, already written and polished, than to hire their own writers to research a topic from scratch. And as the specialization of knowledge has made it harder and harder for a general reporter to get "up to speed" in complex issues of science and technology, the temptation to accept prepackaged stories has only grown. By the mid-1990s, reports *PR Watch*, there were more public relations employees in the United States than news reporters. By 1990, even the *Wall Street Journal* was using news releases as the sole source for more than half of its news stories. The article "Science Has Spoken: Global Warming Is a Myth" was one of them.

A second answer to the question "Who writes these things?" may be that intellectuals often like to be contrarians. The original motive for a writer who falls into a career of picking fights with science may have been little more than the realization that an audacious idea that challenges prevailing thought can gain a lot of attention. For Ben Wattenberg, the realization that some countries' birthrates were falling even as environmentalists continued to talk about population exploding, was no doubt intriguing. And editors and producers often *like* contrarian arguments that come with slick debate and think-tank documentation. If population experts all agree there's a looming problem, an article boldly ridiculing this is bound to ruffle feathers and generate a buzz—and sell a lot of copies to the curious, or to those who enjoy a good fight. All the editor may want to know is: are there enough supporting bits of information available to put together such a story? It's as if the editor were a movie producer evaluating a script for its entertainment potential rather than for its truth. The author is so rewarded for his spunk and becomes so prominently associated with his distinctive view that if later confronted with evidence that clearly refutes it, he may be more inclined to defend his view as though it were his life (which in a sense it is), than to change it.

But it's not just editors who take a special interested in the charismatic contrarian; those who have high financial stakes in what the story claims or denies, too, can make profitable use of a writer who's already on their wavelength. Money and exposure begin to accrue to the writer—there are invitations to speak, to appear on talk shows, to write op-ed pieces. If all goes well, there may be research grants and hefty speaking fees. Before long, the writer has become a kind of information hit man.

News editors and producers will doubtless protest that they do not allow special interests to use their media in this way—that opinions are kept separate from news stories, and are identified as such, and that they are balanced by opposing opinions, etc. But such protestations are disingenuous, because the interests of the media themselves are deeply compromised. To begin with, they are heavily dependent on advertising revenue from the currently dominant industries—the car companies, gasoline retailers, and petroleum-dependent pharmaceutical or cosmetics manufacturers that spend hundreds of millions of dollars on newspaper and TV ads, for example. Beyond that, the media are increasingly owned by larger business entities that have heavy investments in those dominant industries. The major print and broadcast media are directed by executives who are also directors of oil-dependent companies like Ford, General Motors, and General Electric, and who own large blocks of stock in those companies themselves. The prime stockholder of the Associated Press, which controls a major stream of the information reaching industrialized countries, is Merrill Lynch—a company that deals in corporate stocks, and thus profits in direct proportion to the rise of the consumption spike. In the United States, even the supposedly independent Public Broadcasting System (PBS) gets more than 70 percent of its funding from four large oil companies, reports the *Humanist*. So, the news media owe their survival both to their advertisers in the fossil fuel-based economy and to the satisfaction of their corporate managers who are major owners of that same economy.

Many investigators have worried about the growing corporate concentration of media and its potential for abusing power. The specter raised is that the marginalized and disenfranchised peoples of the world will be increasingly ignored by giant entertainment/media conglomerates that have developed a symbiotic relationship with the politicians who get them their licenses or subsidies. This fear is focused mainly, however, on how such growing concentration threatens individual human rights, rather than on how it might threaten the cultural diversity and vitality of humanity as a whole.

In 1995, when Nigeria's military dictatorship publicly hanged environmental activist Ken Saro-Wiwa for having protested against the Shell Oil Company's despoliation of his Ogoni people's land, the outrage of the world was trained on Shell's abuse of the Ogoni, and not on the destruction of their environment (which was the real *cause* of their abuse, and the thing Saro-Wiwa was protesting). Shell learned its lesson: don't blatantly abuse the people, and you can quietly continue to exploit the environment that keeps them alive. In the end, you achieve the same goal (unfettered access to the oil) and with much less interference.

Look more closely at global patterns of abuse, and there may be real reason to wonder whether the endless parade of reports about the misfortunes of individuals—whether they be victims of street crimes, social injustices, or weather disasters—isn't distracting us from a much more pervasive danger, which is the economic system that kills whole communities and cultures. We get very little information about the costs of our consumption in those places that do not yet serve McDonalds or dispense Shell gasoline. Very few of us have ever been informed that of the world's 6,000 human cultures, about 3,000 are nearing extinction.[5] As long as that information can be skewed—through media ownership and the laundering of information so the bulk of news is content empty—we're in danger of being lulled into tragic complacency about our future.

Before the communications revolution, information was not so mediated. In the industrialized countries, the people whose opinions we trusted implicitly were still our parents, relatives, teachers, neighbors, pastors, maybe the family doctor. In the developing world, they may have included the village elders or fellow workers in the fields or fishing boats. Those were people we were physically close to, usually over long periods, so some of the communication came through physical proximity or touch and nonverbal language—strong bonds that go well beyond the logic of words.[6]

In recent years, all of those sources have been increasingly supplanted by remote sources of one kind or another—coming to us not just through news media but through entertainment, commercial advertising, issue advertising, religious proselytiz-

5 ❧ In the late 1980s, anthropologist Jason Clay of the Cambridge, Massachusetts-based NGO Cultural Survival noted, "There have been more . . . extinctions of tribal peoples in this century than in any other in history."

6 ❧ Juliet Schor, author of *The Overspent American: Upscaling, Downshifting, and the New Consumer,* describes this disconnection as a shift away from "proximate reference groups"—a tendency for Americans to form their values, expectations, and aspirations from the fictional shows they see on TV, rather than from what they experience in their real homes and communities.

ing (much of it via TV), and corporate PR. George Gerbner, a former dean of the University of Pennsylvania's Annenberg School of Communications, has calculated that the average American child sees 40,000 murders on TV by age 18—without having to experience, or therefore begin to comprehend, the personal impact of any of them on anyone. Gerber, who fled to the United States to escape the fascists in World War II, describes this desocializing effect of TV as another form of fascism. Similarly, it has been estimated that the average American sees 150,000 commercial advertisements on TV during his or her lifetime—most of them very disarming and entertaining invitations to more consumption offered not by parents or friends who have the viewer's wellbeing in mind, but by corporations that are often as large as countries. So, our acquisition of information is increasingly filtered through large organizations—whether they be the corporate owners of newspapers or radio stations, their advertisers, or the industry lobbyists who help set the rules (such as the allocations of bandwidths to commercial broadcasters) by which the media draw their profits. Most of us have no idea where those organizations are physically located (if they can be said to be located anywhere at all), or where their key players live.

That the information we get is heavily filtered is nothing new, of course. Media-savvy consumers learn to be both skeptical about what they hear and cynical about the motives of the people or institutions doing the mediating. At the same time, both reporters and their readers seem fascinated by the battle to control the world's information technologies, infrastructures, and organizations. This fascination is much like the popular culture's fascination with spectator sports. Sports are ritualized conflicts, and the majority of people seem more inclined to get excited about these ritualized battles than about real ones. That may be a good development if it means replacing bombs with soccer balls, but problematical if it means ignoring of real threats. In 1998, for example, millions of Brazilians would have told you the most terrible thing that had happened in their country that year was the nation's crushing defeat by France in the World Cup soccer championship. Few, apparently, would have said it was the government's decision to suspend the country's environmental laws for the next 10 years—a decision that opens the Amazon to even faster slash-and-burn destruction than it has suffered so far.

What's most amazing about the amazing phenomenon of the communications revolution, then, is not that Leonardo da Vinci's paintings could all be stored on a sliver of silicon in your pocket, it's that we've become so fascinated and obsessed by the expanding capabilities of this revolution and its celebrity performers that we have virtually lost interest in what information is being transmitted and how it is being used to shape the world's future.

In this respect, there's a striking similarity between our consumption of information and our consumption of food. Most of the food products sold in supermarkets today contain ingredients that are not there for reasons of nutrition or health, or in some cases even for taste. They're put there to fatten livestock, make fruit glisten, provide cheap fillers, extend shelf life, make oils stiffen like butter, kill insects, or help us get addicted. Some of these ingredients have been found to contribute to cancer, heart disease, or hypoglycemia, not to mention obesity and teenage zits, but they're used because they pump up manufacturers' or retailers' profits. Yet, most people are apparently uninterested in the ingredients of their food (except perhaps in whether it contains the demons "fat" or "cholesterol") and wouldn't think of asking in a restaurant, for example, whether the chicken is fried in unsaturated vegetable oil or hydrogenated fat. There is an implicit faith that if a product is for sale, it's safe. A similar faith pervades our acceptance of information. We know there are a few people in public life who tell lies, but these are scandalous or criminal exceptions to the general rule that if it's printed, there must be something to it. Sex, violence, and profanity are the fat, cholesterol, and sodium of information: we say we want them labeled, but we keep consuming them without restraint.

All the major information services have their additives—their artificial stimulants and other ingredients designed to addict, satiate, or manipulate. To assess what this does to our ability to recognize and assimilate real knowledge without its being hopelessly adulterated, it's essential to identify what the additives are, and how they can alter our perceptions. In other words, if our perceptions of the biophysical world are *already* distorted by marginalization, specialization, foreshadowing, and systematic disappearance as described earlier, how are these further distorted by the inherent characteristics of the major information sources? All of these sources—our entertainment, advertising, preaching, and PR, as well as our conventional news media—have biases that systematically skew their selection of what information to impart, how much context to provide, how long a view to take, and what spin to put on the message. Moreover, all but the religious category have essentially the same biases. All emphasize

- the *sensational* over the *analytical* (excitement over thought)
- the *sudden* over the *gradual* (incidents over trends)
- the *immediate* over the *long-term* (the present over a decade or a century from now)
- the *narrow* view over the *broad* (expertise over perspective)

- *growth* over *stability* (youth over experience)
- *consumption* over *sustainability* (gratification over health)
- the *temporary* over the *permanent* (liquidation over reinvestment).

Among the main categories of information sources, commercial news is by no means the largest in total volume of information in most countries, and for many people it may be one of the smallest. But since it's the one that seems most widely presumed to be commited to an "objective" picture of events, it warrants a closer look. How does the news rate on these biases?

In all news media, the competition for audience skews selection of content in favor of the sensational ("If it bleeds, it leads") and the sudden (to get yesterday's audience back today, you have to show something that happened last night).[7] The quick

7 ❧ Historians a century or two from now may look back at the last months of 1997 as a time of incipient change more awesome than any other in the human record; that October, November, and December had been the warmest of those respective months ever recorded—even as the climate treaty conference took place in Kyoto; and the following year, in addition to extending that escalation of global warming month after month, would bring the most sudden manifestations of climate change since the end of the last Ice Age. Scientists pointed out not only that the climate was changing with a speed that demanded urgent action at every level of human life, but that a biological crash more precipitous than any since the dinosaurs were extinguished was now underway. Yet, the disconnect between the momentousness of these events and the myopia of the media was so profound that Clarence Page, a reporter for the *Chicago Tribune,* when asked to explain the "emptiness of the news" as the year approached its end, told a Washington, DC radio station that, frankly, "nothing was happening."

pace of news production favors the narrow over the broad and the immediate implications over the long-term, in part because that's what shortening reader attention spans demand, and in part because that famously fast pace (a whole new set of news each day, and at least some new events or angles each hour) doesn't give reporters time to look any further. The shortening of attention spans is caused in part by these tendencies toward quicker, narrower, more sensational coverage to *begin* with, and further exacerbated by the interruptions of commercials, which themselves put an even greater premium on the use of quick impressions. Thus, the news medium feeds a vicious cycle: the more successfully a 15-second message can manipulate your perceptions, beliefs, or desires, the higher the price it brings (whether as news or as advertising), and as the price gets higher, the message carrier has to perform that manipulation even more intensively, or competitively, in order to achieve the behavioral responses—your purchasing, voting, sending money, or turning on that channel again—required to justify the price of production.

But the shortening of attention is also driven by the acceleration of change, because if the perception of time is affected by the amount of change, then something ten years in the future is as difficult to think about or plan for now as something a century or two in the future may once have been. Why plan for what you can't anticipate? This inclination to focus on living now, not later, provides an accommodating medium for advertising, promotion, and PR, and for the high consumption they encourage.

Now, consider how this pervasive bias acts on our perception of the four spikes and their interactions. The illustrations that follow are from recent years (or days, as this is written), but the past is the bow that looses the arrow of the future. You can be sure of this: extreme weather is coming to your area soon; strange insects or diseases are coming too; more people are coming; and either more shortages of water and other resources or painfully higher prices, if you're spared the shortages, are coming as well. The question is whether your message carriers will manage to skew the information so that you don't realize its immensity until it's too late to respond optimally—or even to respond at all. I say "manage," because while some of the skewing is inherent in the biases described above, much of it results from a premeditated exploitation of those biases. Here's how it works.

- *Is the news of the spikes sensational?* No. It's about an invisible gas that's always been in the air and that is harmless to breathe. An unobservable disappearance of plants and animals, the majority of which we've never seen to begin with. An increase in population that's visible only in the form of a gradually mounting annoyance about other issues that are connected to it (local "slow-growth" vs. "pro-development," money for roadbuilding vs. money for schools, etc.), but which are discussed only as isolated issues. As for heavy consumption, it's visible enough, but so pervasive in rich countries that it's seen as a entirely normal. So, when media dwell on crimes, crashes, and scandals, they provide a rich diet of distractions from news of the spikes that are killing us. For those who have interests in keeping the public thus distracted, no conspiracy is necessary; all that's needed is to let the media do what they already have a strong financial incentive to do.[8]

- *Is the news of the spikes sudden?* Ironically, no. If we don't act soon, the irony will be seen by historians a century or two hence as excruciating and colossal. After all, as observed earlier, from the standpoint of geological or evolutionary time, these are extremely sudden. But from the enormously contracted viewpoint of daily and hourly newsgathering (which grows still more myopic as change accelerates), they are only gradual trends. In the news editor's view, there's no perceptibly greater reason to report the CO_2 concentration or human population today than there was yesterday. The liquidation of the planet's resources is like the use of gas in a car, which has become perhaps the most paradigmal form of consumption: a car goes just as fast when its tank is only one-eighth full as when it was full. If you don't pay attention to the gauge, you may drive blithely along until the car stops. The most critical case of such full-until-empty orientation is the consumption of fresh water, which in much of the

8 ❧ Anne Platt McGinn, a researcher at the Worldwatch Institute, released a report on ocean fisheries, describing how overfishing had begun decimating not only the adult fish populations in most major fishing grounds, but the juveniles as well—thus accelerating the unraveling of the oceanic food web. A reporter for the *Toronto Star* wrote a story on the report that was to be published in the next day's edition. But the next morning, the reporter called McGinn to say there had been a sensational late-breaking story, for which the editors said they needed to make space: a member of the Spice Girls, a pop singing group, had announced she was leaving the group. Readers would want to know all the details. McGinn's story would, regretfully, have to be cut.

world now exceeds rainfall recharge. Anywhere that it does, the water tables are falling. Yet, until they run out, users can continue to consume (in that area, at least) at accustomed rates—oblivious to the coming shock of hitting bottom. (Like a gas tank, a nonfossil aquifer can be replenished, but once it hits bottom the level of use has to be cut abruptly to the rate of recharge.)

- *Do the spikes affect short-term interests?* Here the answer gets tricky, because the spikes can generate both long-term and very short-term secondary effects. But rising carbon gas as a primary phenomenon (as a phenomenon of increased solar heat-trapping and rising ground or air temperature) might go on for decades longer before it's much felt in some regions, so the carbon spike might not much affect, say, the short-term prospects of an investor in ski resorts who's looking at the average temperature for Pennsylvania and can shrug off a 1- or 2-degree increase from global warming during the depreciation period. But the secondary effects of global warming, such as big storms, floods, droughts, or infestations, are a different story. There the effects are immediate and do get major news coverage, but that's also where the disconnect between short-term and long-term implications kicks in, as happened in Red River. The news team focuses on the raging waters, the damage, the emergency shelters, the receding waters—and then perhaps revisits briefly for the rebuilding, though usually it does not. The longer-term issues are crowded out by the demand for new events in tomorrow's news, the lack of journalists' training in the interconnectedness of events, and their lack of research time in a hectic, always-accelerating, business. And, as further discussed below, all the financial incentives are geared toward satisfying short-term interests in the coverage.

- *Are the spikes conducive to narrow focus?* Here's perhaps where the disconnect is greatest. It is, in part, *because* the megaphenomena are so pervasive that they are so hard to treat as news. When they appear in the news at all, it is usually in tiny pieces—focusing on a detail here and a detail there, but never drawing any connection between them. For example, consider three widely separated recent stories: (1) The *Australian Finance Review* publishes an article about the Freeport McMoRan gold and copper mine in Indonesia, noting that the huge mine dumps 160,000 tons of toxic waste into the Ajkwa River each day. The local communities of the Amungme and Komoro people say the waste is killing the fish on which they depend for a living. The president of the mining company, an American named Jim Bob Moffett, is quoted as scornfully replying that the environmental impact of the mine is "equivalent to me pissing in the Arafura Sea"; (2) In the southern United States, where wood pulp (for making paper or packaging) is a major industry, the Alabama *Mobile Register* quotes a University of Alabama professor, Robert Lawton, as saying "We've got another 20 to 40 years of pine cropping before we are blown out of the water by cheap cropping in the tropics. . . . In the game of low-quality, high-volume production, Alabama is gonna get its pants beaten off"; (3) in Japan, *Yomiuri Shimbun* runs an article describing the attempts of fruit growers to eradicate macaques, a type of monkey that has taken to raiding their orchards. The first of these stories seems to be about a conflict between mining and indigenous rights; the second about tree plantations and the pitfalls of global commerce; and the third about farmers' struggles with pests. There is no occasion, in any of these publications, for noting how such disparate stories are connected. Yet, there is an enormity here. The poisoning of rivers, the replacement of natural forests with monoculture plantations, and the eradication of animals whose habitat is now claimed for human activities—all global phenomena, repeated in thousands of localities and reported in thousands of narrowly focused stories—add up to the extinctions spike. But rarely is there any news of the spike itself.

- *Do the stories of the spikes satisfy the prevailing public appetite for growth and consumption, as opposed to stabilization and sustainability?* The answer might appear to be that they do, on an epic scale, except that what they mean is hugely misconstrued in the telling. Because economic growth is widely seen as a positive phenomenon, and as one that can continue indefinitely, the indicators of growth are treated in a good-news-is-no-news fashion. The perception of growth is truncated by the short-term perspective —the focus on its immediate rewards.

That's the news. Of the other main categories of information sources, at least five are even more blatant in their catering to these biases toward the sudden, the short-term, the narrow, and the consumptive.

Advertising, by its very nature, generally has no purpose other than to maximize consumption in the short term, with as narrow a focus on the advertised product as possible.[9] Advertising has also played a leading role in developing the use of intensified sensation to attract attention, causing a progressive reconditioning of brain synapses analogous to that induced by narcotic drugs, in which as rival products compete for attention the advertiser has to become still more sensational to stand out, and as viewers become jaded, the stakes are raised still more.

Entertainment, like advertising, tends to shut out awareness of anything outside its own performances. While ads do this by focusing on a particular product and its satisfactions without regard for their consequences (whether those consequences be obesity, cancer, or climate change), entertainment does it by *becoming* the product that satisfies and shuts out all other awareness. The two overlap increasingly (entertainment programs are filled with cross-promotions) and are sometimes almost indistinguishable.

Corporate PR is only slightly less narrowly focused than product advertising and more dangerously deceptive. The danger lies in the fact that some corporations are now as large as nations and have comparable power, yet are even less accountable. Well aware that publics are wary of such unchecked power, corporations have used PR to defuse potential conflicts with the public over their impacts on public interests, and in doing so have leaned over backward to act as though they are global nations interested in "serving" the public responsibly. This effectively

9 ❧ It may have innocent origins in simply alerting potential customers to the availability of a product they already wanted, as in a sign that hung from a house saying "Inn." But there's a fundamental difference between facilitating the satisfaction of an existing demand and creating a new demand where no real need existed. In a world of overpopulation and overconsumption, that difference is a moral one.

draws attention from their intention to sell as much product at as high a profit margin as they can, which in the prevailing economy often means liquidating public assets. Not all companies play this game, and many corporate products not only serve public interests well but will be essential to any hopes of arresting the spikes. Corporate PR could be turned to facilitating that effort, but so far has worked mainly against it.

One reason corporate PR has been particularly effective in pushing the world's most momentous issues to the margins of public consciousness is that it often appears to have no moral keel. It has no rules of conduct that tell it to promote the company or industry within a framework of fundamental respect for (1) the integrity of the environment on which that company or industry relies for resources and waste disposal, (2) the society on which it depends for a market, or (3) the language with which it conveys its messages. The nuclear industry, for example, often advertises its benefits for clean air (because it doesn't emit smoke), while not mentioning the radioactive holocaust of soil and water it wreaks when nuclear waste is leaked or dumped—as has happened in hundreds of sites worldwide. (The industry's latest slogan—"Nuclear, more than you ever imagined"—is more appropriate than its purveyors may have intended.) Similarly, the timber industry advertises a deep caring for wildlife, but in view of the destruction its operations have inflicted on habitat, it is clear that that industry is far more interested in what kind of impression it leaves on consumers than on what kind of footprint it leaves on the planet.

But perhaps the most troubling trend has been the perfection of a technique George Orwell warned of in his essay "Politics and the English Language"—that of using words to mean the opposite of what they have been generally understood to mean, thereby robbing them of their power and disarming those for whom the original meaning of the words was important. Many companies have done this simply by relabeling their environmentally destructive activities as environmentally friendly. The advertising slogan of Weyerhauser, a company that is in the business of turning forests to pulp, is "Weyerhauser, where the future is growing." And then there is the work of E. Bruce Harrison, who now works as a PR consultant. Harrison has written a book, *Going Green: How to Communicate Your Company's Environmental Commitment.* The book isn't about how companies can clean up their pollution, but how they can clean up their images—as though it isn't the pollution itself, but only the perception of it, that is a problem. As it happens, this is the same E. Bruce Harrison who led the chemical industry's 1960s disinformation campaign against Rachel Carson's *Silent Spring.*

Political ads and PR may be a little less dangerous than the corporate variety, if only because their budgets are smaller (they're really just one part of the corporate budget), but they're just as deceptive. Moreover, the performers who present them often seem quite oblivious to the real implications of what they're saying. An American politician, for example, will speak out against abortion in a voice that shakes with emotion, because he's deeply "pro-life"—yet will just as resoundingly promote such life-extinguishing measures as capital punishment, unrestricted sales of guns, and the criminalization of abortions that would save mothers' lives. Less obviously, the pro-life politician will often support "pro-development" policies that are killing off the planet's life at a staggering rate. By separating issues that nature can't separate, political ads act like mental hand grenades thrown into intelligent discourse. With their extremely short-term missions of winning elections or legislative battles, they are inherently blinding to public perceptions of global change.

Conscientious and enterprising reporters occasionally break through the constraints of their profession and give us fleeting glimpses of the spikes themselves. That happened in the summer of 1998, for example, when the *New York Times* published a special news section on extinctions. And there's a subculture of people, communicating by e-mail and other means that bypass mass media, who are alerted and trying to mobilize. But even the few reporters who do manage to break through are unable—without systematic massive support from the media organizations and public—to track the *interactions* of the spikes. The *New York Times* section on extinctions, for example, was separated from the rest of that issue as though it had nothing to do with the "main" news appearing in other sections. On the front page of that day's paper, an article on California's primary election campaigns noted that with the economy fading as an issue, politicians were "struggling for something to talk about." Few readers would likely have experienced any sense of irony in reading that observation in the same issue that carried a special report on what may prove to be the single most momentous development in the history of the world. Nor was any connection made between the content of the special extinctions section and the coverage, on other pages, of the unprecedented weather disasters happening around the country that summer.

The greatest danger in our information delivery, of course, lies not in the failures that are inherent in our media, great as they are, but in the deliberate exploitation of those failures by organized interests. When the fossil fuel industry was confronted with the prospect of a global climate treaty in 1997, its bottom-line goal was to assure that the United States, which is its geographical base of operations, largest national market, and key to the global market, would never actually ratify a strong climate treaty. The Global Climate Coalition (GCC) made a big show of being worried about Kyoto (hence the *Washington Post* ad in which 160 major corporations urged the US government not to act in haste), but of course the agenda in Kyoto was only to approve the wording of a treaty, not to secure the actual ratification (approval by the various countries' legislative bodies) that would put the treaty in force. The ratification was supposed to come a year later. But by protesting loudly, the coalition was able to accomplish two strategic purposes. First, after the United States signed at Kyoto, climate scientists and environmentalists would smell victory and would stop pushing so hard. Second, while the activists trooped off to fight some other battle, the US legislators would be sure to remember the real message: that whether or not they supported signing at Kyoto in order to gain approval from a vaguely concerned segment of the public, they dared *not*, if they valued their jobs or their biggest contributors' continuing support, think of ratifying anything that could "ruin" the US and global economy. It was a deftly framed threat, because it did not make a *case* that controlling emissions would be economically ruinous—it treated that as a given.

In other words, the GCC was too smart to get caught offering politicians a bribe. Americans generally accept the theory that their lawmakers can be bought, and it would not go well for oil and coal companies if the public thought they were bribing politicians in callous disregard for the health of the public. Better to let the public think the politicians had supported the accord despite the protests of the oil and coal interests—and then, later, had on their own seen the wisdom of refusing ratification because of the economic damage that would result.

The trick was to get legislators to accept the economic concern as a given, and the way to do that was to create an impression that that issue had already been decided. Then, any politician who asked "how do we *know* this will hurt economically?" would look foolishly uninformed.

The GCC's strategy, then, was to plant the seeds of this idea of the treaty as "economically damaging" as profusely as possible, concentrating them not directly on news reporters and analysts, who were too likely to question the source, but on sources the reporters and public would be more likely to trust. As noted earlier, there are always a few experts who can be recruited to argue any cause, if the rewards—in money, notoriety, or intellectual "vindication"—are tempting enough. The collaborators were recruited, and their testimony was packaged as the "findings" of a phalanx of scientific organizations or journals, whose considered opinions were then reproduced by the millions and dispersed to message-carriers as mainstream science.

It worked like this: A mainstream message carrier (news reporter, politician's speechwriter, PR strategist), wavering on what position to take on this troublesome climate issue, would receive an urgent letter—often hand-delivered, often from someone whose title would seem vaguely familiar. Inside would be a summary of a report from what appeared to be a respected scientific organization or journal, often replete with impressive looking graphs and tables. The cover letter would be quite frank: it would urge the recipient to oppose the climate treaty because of the dire effect it would have on the economy, and it would imply that the enclosed technical material—if the recipient wished to study it—was the proof. Millions of such messages went out in the months and days before Kyoto.

Few reporters, speechwriters, or other opinion-leaders read—

or perhaps were even able to read—the "technical" material. If they had, they'd have found it was bluff: there was no proof at all—no evidence of damage to the economy. But there was always a well-crafted press release enclosed, which provided vivid material for a news story, speech, or policy paper. Among the urgent messages thus circulated were these:

- A press release from "The Advancement of Sound Science Coalition" dated December 3, 1997, saying, "Given that economic prosperity is necessary for a cleaner environment and improved public health, *action that harms the global and national economies should not be taken without compelling evidence of need.*" (Emphasis added.) The release contained no evidence that the treaty would harm economies—the suggestion being that everyone knew that already. A few months later, this would be echoed in the Exxon Corporation shareholders' magazine *The Lamp,* which argued that when it comes to climate change, "before you act, you first find out whether there's a problem." The magazine suggested that another 10 or 20 years of research is needed before we'll know whether we need to take action. "The trick is not going to be reducing emissions, but reducing them without wrecking the world economy," it warned.
- A news release from the Competitive Enterprise Institute (CEI), a group that identifies itself as "dedicated to free markets and limited government." The release bore a boldface head: "White House Fabricates Greenhouse Consensus." The headline, by itself, was an Orwellian master stroke, since it was organizations like the CEI, the intellectual commando units for the Global Climate

Coalition, who were, in the act of writing that very headline, doing the fabrication. The consensus identified by the White House was the real one. The CEI release went on: "It is irresponsible for the President of the United States to . . . lock the US into a binding UN treaty that will cripple our economy." Again, there was no evidence enclosed that the treaty would cripple the economy. (The use of "binding" gives away the CEI's real intention, which was aimed not at stopping the signing of a treaty, but at stopping *ratification* of anything truly effective.)

- A release from the "American Council on Science and Health" stating, "Many debates have erupted over global warming and the most disturbing one concerns the apocalyptic predictions of the effects of global climate change on human health." This sounded a theme that recurred again and again in the campaign: the reference to climate scientists' projections as "apocalyptic."
- A letter to selected media and policymakers from Frederick Seitz, a past president of the National Academy of Sciences, urging the recipient to read a technical paper from the Oregon Institute of Sciences and Medicine—an article that looked exactly like a reprint from a scientific or medical journal, though on close examination it was not. This piece took a different tack. Seitz wrote: "This treaty is based upon flawed ideas. Research data on climate change do not show that human use of hydrocarbons is harmful. To the contrary, there is good evidence that increased atmospheric carbon dioxide is environmentally helpful." Around the time this appeared, GCC spokesmen appeared on talk shows and pointed out that CO_2 is in fact a "natural" substance, essential to the growth of trees. It was like

> arguing that taking overdoses of sleeping pills is not something to be concerned about (or to take precautionary measures against), because sleeping pills help people to rest. Shortly after the "Oregon Institute" letter was distributed, it reappeared in the form of the *Wall Street Journal* article mentioned earlier, "Science Has Spoken: Global Warming is a Myth."

Thus these ersatz "news" releases and "scientific reports" were not seen by the general public, but were digested by the mainstream information carriers and reissued in popular formats, almost never with any explanation of where the idea that the treaty would "cripple the economy" had come from. What the journalists wrote was that because of the economic burden, a group of scientists (or doctors, or businessmen) were opposed to any US action on the treaty. Left largely unnoticed, in this process by which the attention of an overwhelmed reporter is grabbed and commandeered, was a statement that had been issued earlier that year by more than 2,500 economists—led by Nobel laureates Kenneth Arrow of Stanford University and Robert Solow of the Massachusetts Institute of Technology—stating that the United States would be able to reduce its industrial emissions enough to slow climate change without damaging its economy.

The process by which the GCC's positions were presented first as presumed scientific or academic sources (with no mention of fossil fuel sponsors), and those sources were in turn repackaged as mainstream news media commentaries, constituted a form of information laundering, in much the same way that some Miami or Nassau bankers operate as launderers for drug money. While the GCC corporations couldn't risk being too obvious about buying senators' votes directly, through overly fat campaign contributions, they could easily buy a few academics who were starved for research funds or attention. Thus, some of the source organizations were front groups set up by the GCC for the explicit purpose of protecting fossil fuel interests, and once the treaty was gutted those organizations disappeared. The ideas they had planted, however, spread persistently.

The Senate voted unanimously to refuse ratification of the treaty as framed in Kyoto. And when the delegates reconvened a year later in Buenos Aires, nothing had changed—except the climate.

6

INFORMATION DENIED, DISREGARDED, OR DISAPPEARED

ONE DAY in 1987, near the Brazilian city of Goiânia, two boys found an abandoned piece of equipment. It was boxlike and mysterious, and as boys will, they broke into it. Inside they found a glowing blue powder. One of the boys put some of it on his face like war paint, and grinned garishly. The other put some in his hair. They laughed, threw the powder at each other and checked out whether it had any taste. Then they filled their pockets with the wondrous stuff and took it home to show their friends and siblings.

Authorities later determined that what the boys had found was cesium-137, a highly radioactive substance. A few days later, the two companions, along with several of their family members and neighbors, had to be buried in lead-lined coffins.

The story of those boys could be seen as a parable of the human infatuation with technology, except that the larger story—*our* story—has not yet been completed, and has a range of possible endings from which we now must choose. As a species, we've experienced the same curiosity, pleasure of discovery, and impulse to play with things we don't fully understand—or in some cases don't understand at all. We love the adventurousness of play—the thrill of discovering some new way of extending our own powers beyond anything we had imagined possible. Whatever early hominid discovered how a log could be used as a lever to multiply the power of his arm several-fold, and enable him to move a boulder, must have felt a tremor of excitement that was to become selectively strengthened over the generations. The Brazilian boys, as modern Homo sapiens, were genetically predisposed to love the way they could refract the light on their faces into a supernatural blue in a way they had never been able with mere clay or paint.

The fascination with technology drove the Industrial Revolution, and public policy today is still imbued with it. The issues we debate still focus on how we will manage our technological powers: our communications infrastructure, genetic engineering, highway and bridge construction, and weapons design. Our policymakers are more interested in who will control the biotech patents, the broadcast frequencies, and the industrial oligopolies, than in what they will use them for, and why.

In the future, policy may have to be more concerned with psychology. There are mysteries about human behavior that are becoming matters of increasing urgency, yet seem to be of little interest in the dominant institutions of business and government. The people running those institutions seem to regard these mysteries mainly as appropriate themes for entertainment, or perhaps for volunteer work, but not for real work.

Many industries take it for granted, for example, that there's money to be made from human addictions (to sugar, drugs, alcohol, caffein, tobacco, sex, gambling, TV, or even shopping), but few have any interest in what addiction is, or how it works, except to learn how it might be made still more profitable. *Why* do people become addicted, by the tens of millions? Why do individuals, and sometimes entire societies, engage in self-destructive behavior, sometimes with little or no apparent effort to change course? We can peer back through the mists of time to the Easter Island or Mesopotamian or Mesoamerican cultures and wonder how they could have failed to see the slippery slope and to try to arrest their falls, but paradoxically we can't see our own addictions and self-destructive behavior, which are far more blatant—and which increasingly encompass the whole planet. Unlike earlier cultures that at least had a chance to move on to new territory (even the Easter Islanders weren't trapped until they'd cut down their last large trees and eliminated their ocean-canoe option), our global culture has nowhere else to go. Global warming and bioextinctions are all-encompassing, and the expansions of population and consumption trigger chain reactions that cannot be stopped by shores or borders. Yet, even when the information we need in order to respond is not being distorted by marginalization, or put beyond our comprehension by specialization, huge amounts of it are kept out of our awareness altogether. Some is blanked out by personal perceptions, some by the deceptions of the dominant economic system, and some by systematic hiding.

1. DENIAL

One answer to these questions of mass addictive and self-destructive behavior is that a staggering number of us have fled or slipped into denial; we don't *try* to stop the addictive behavior because even when the barriers of obfuscation are pointed out to us—even when we have the scales removed from our eyes—we still don't truly believe we're putting our whole world at risk.

Why we don't believe that, when our leading scientists do, may be the most important question science faces today—more important than whether disease can be conquered by genetic engineering, or whether we will find higher forms of life outside the solar system. The question of denial has primacy, because denial is what blocks our ability to make progress on all other fronts. It's what locks us into those patterns of addiction and self-destruction, which become more menacing the longer we remain locked in. Denial is the flip side of sentience, and sentience is what has to separate us from the ants that sink with the log.

The pervasive information laundering described earlier, which washes the content out of most messages and leaves them stimulating us like the mental equivalents of empty calories, is dangerous enough in its own right. Even those who *know* the manipulation is happening have to work hard to see the process with any kind of clarity—and they may have to do frequent reality checks to assure themselves that they are not delusional.

The few who do this are able to do so by monitoring the most believable of unlaundered information sources, such as the IPCC climate studies, the IUCN study of extinctions, or the UN population projections, as well as the peer-reviewed articles published by scientific and medical journals. Most people don't have access to these sources (or don't have any way of knowing which of them are credible) and live in a world where the vast bulk of information is suspect. In such a world—where the majority already have only a tenuous grasp of what's happening to them—the power of denial can be fatal.

Among people as diverse as religious leaders and scientists, there's a surprising convergence of views that humanity has reached a critical point—a decision point, at which we either consciously lift ourselves to a higher level of sentience or fall back and decline as a species in the way many of our individual civilizations have declined in the past. In the language of evolution, we now have to adapt extremely rapidly to the changes we have wrought—our adaptation necessarily coming as quickly as the changes we have made in our environment. In the language of religion, God has given us an offer: to see the consequences of our actions and assume moral responsibility for them, or to be consumed by them.

The consciousness that distinguishes Homo sapiens from other animals (though perhaps some other species are more sapient than we yet realize) can be more specifically characterized as the ability to conceptualize, imagine, or foresee, as well as to remember—in other words to have mental experience outside the immediate physical environment. What makes us human is the ability, then, to see from the eyes of someone in the past or in the future, or in another place, or someone who has different tools or abilities than the ones we have at the moment. In other words, to be human is to be able to envision, as well as to see literally, and to empathize, as well as to feel one's own needs.

What addiction does is to cut off this envisioning and empathizing—to cut it short and keep the addicted person focused on immediate experience. That experience may be intoxication, a drug high, sexual gratification, or TV. It may also be the satisfaction of a compulsion to shop, to collect fad dolls, eat sugar, or gamble. It cuts attention off from long-term consequences—from the risk of accident, obesity, financial disaster, sexually transmitted diseases, ruined relationships, a dulled imagination, or a ruined agricultural base. So, one answer to the question of why people become self-destructive is that when they give up the capacities to envision and empathize, they also give up the remarkable ability to *survive* that our evolving consciousness gave us.

Environmentalists have often noted that industrialized society is "addicted to oil" or "addicted to cars" or even "addicted to shopping"—to excessive consumption. This use of the term is meant to be metaphorical. But if we define addiction not in terms of as-yet-not-understood mechanisms of the brain, but as patterns of behavior that cut off empathy and envisioning—then these addictions to cars, oil, and shopping are real. The appetite for gasoline that allows industrial nations to support an industry now known to universally raise the risk of jeopardizing civilization itself—yet allows us to do this without envisioning anything at all beyond the gratification of pulling away from the pump with a full tank—is addictive on a societal scale.

The stupifying fact that tens of millions of people have actually chosen to deny themselves the experience (and accompanying responsibility) of looking back and ahead, in so much of their activity, means—regardless of how it came about—that a huge market for denial has developed. Like any major market, it is segmented. Different people want to close off different rooms in their consciousness. So, there are some fairly "targeted" forms of denial available: a bout of drinking to forget a bad day, or a book that expunges our guilt by assuring us our unhappiness was our parents' fault, not ours. There are also some broader forms, however—aimed at relieving us of any worries about the future at all.

Fundamentalist religions, while condemning immoral behavior, also assure their believers that whatever happens is destined by God—effectively relieving them of responsibility for any effect they may have on their own destiny. If Armageddon comes, it is God's will, and the implicit consequence is that we don't have to prevent Armageddon—and indeed cannot prevent it. Similarly, a doctrine that tells people authoritatively what is right or wrong, and what is good or evil, without asking them to exercise their own judgment, effectively relieves them of responsibility for performing one of the primary tasks of a sentient human—to participate actively in the search for understanding.[1]

The world's churches are now deeply divided on the issue of human participation in the shaping of our species' destiny: while many fundamentalists say it is not for us to question God's intent or authority, other religious leaders say questioning is *the way* to come closer to God.

The technological optimists, at first take, are diametrically different from the religious fundamentalists. These are often high-profile opinion makers or financial moguls. But under their worldly surface, they're much the same: they are driven by faith rather than curiosity.[2] Many are people who, having "made it," perhaps want to secure their positions by believing comfortably rather than having to be forever learning. School was a means to an end, and they shouldn't ever again have to be annoyed by teachers. Their faith is not in the Bible, but in what got them to where they are—technology. Bill Gates, whose net worth rose past $50 billion in 1998—making him 50,000 times a millionaire—might seem to be the ultimate techno-optimist. Gates envisions a magnificent future for all of us and has gone to great pains to give us a preview of it in the design of his now legendary

1 ❧ There's an analogy between public wellbeing and personal medicine; individuals who want a high level of health, as opposed to just an absence of debilitating illness, have learned in recent years to participate actively in their own care, and not just defer passively to what the doctor says. It is this that has pushed medicine more than anything else.

2 ❧ In this, there's a great chasm between science and technology, which are often assumed to be two sides of the same golden coin. The scientist is by nature a questioner; the apostle of technology brims with certainty.

house on Lake Washington. All he needs to know is what is technically possible, and of course everything in his house is possible. In the techno-optimist view, the 4 million squatters in the shantytowns of Cairo could someday *each* have a mansion with a library that senses their entrance and plays their favorite opera music and does the same for each of their house guests in their individual guest rooms. But Gates can entertain this generous fantasy because he doesn't have to account for the costs. It's not that he doesn't have to count the purchase costs to the squatters; of course he knows that has to be paid, but if he was able to increase *his* wealth 1-million fold over a decade, in theory they can too. In conventional accounting, however, there's no requirement to count costs to the ecosystems that will provide all these newly rich Egyptians with their building materials and energy supply and fresh water.

But wealth aside, Gates and his cyber-brethren differ in an important way from the people who make up the hard-core techno-optimist market. Gates and company are dreamers; they do envision. The trouble is, they envision with laser-narrowness, speeding forward with uncanny consciousness while oblivious to what is approaching from left or right or underneath. They are childlike in their confidence—like a child chasing a ball into a street. When the spikes begin to pull down the web of life, the World Wide Web will come down with it. For the first time in their lives, perhaps, the young moguls of high tech will look up at the world around them, and they will be amazed.

More dangerous than Gates and his fellow laser-seers are the many less well-known but still influential people who see what technological innovators have done, are dazzled by it, and have been persuaded by it that if that can be done—well, "we" can do anything. But "anything" is, for them, a blank slate. They do not envision solutions themselves, but are confident someone else will. Often, this confidence comes across not as quiet assurance, but as an aggressive, excited insistence that you, too, become a believer.

A particularly breathless example of such insistence appeared in a cover story of *Wired* magazine in the summer of 1998. It was titled "The Long Boom: a History of the Future, 1980-2020." Under that was this in-your-face subtitle:

> We're facing 25 years of prosperity, freedom, and a better environment for the whole world. You got a problem with that?

The apparent implication was that if you didn't agree with what followed, you were against prosperity, freedom, and a better environment.

This *Wired* article, unlike the *New York Times Magazine* piece on population (which drew dangerous inferences but at least drew them from real facts), contained no actual analysis and consisted almost entirely of adolescent daydreams about a future in which there are no problems and everyone is incredibly rich. A few examples of its technique suffice to show how this sort of fantasy is promulgated.

"We have entered a period of sustained growth that could eventually double the world's economy and bring increasing prosperity for—quite literally—billions of people on the planet," write authors Peter Schwartz and Peter Leyden. To a reader yearning for good news, this may sound promising indeed. The trouble is, for the world economy to double would be neither good nor news. The authors seem unaware that the global economy has already been busy doubling and doubling again for the past half-century (from $4.9 trillion in 1950 to around $36 trillion in 1998), as the graph of the consumption spike shows; yet the gap between rich and poor has only widened in that time.[3] The "increasing prosperity" the authors expect from such growth is nothing more than a repackaged version of the long-discredited trickle-down theory of the 1980s.

The *Wired* article encourages us to put our faith in a very appealing but nicely vague concept of "interconnection" (as in the notion that we're now all wired together by global telecommunications and the Internet). Yet, the kind of global "virtual" community this entails is being achieved by an accelerating *dis*connection between us and the real natural world that supports us. If conservatives in denial are inclined to retreat to gated enclaves or guarded office suites, others may find reassurance in the idea of a completely wired world in which we have, as this article puts it, "Everybody on the planet in the same economy." The authors avoid alarming the reader with any mention that it is this very movement toward having everyone in the "same economy"—the same McDonalds in every Asian and African village as in Denver or Dublin—that has destroyed genetic variety in potatoes, cultivated pesticide-resistant insects, left hundreds of millions of farmers without land or livelihood, and impelled Chinese women to swallow pesticides in despair. They also avoid

3 ❧ A few years ago, the United Nations Development Programme (UNDP) did a survey of human progress over the past half-century and reported that between 1960 and 1990, the income gap between rich and poor had doubled, worldwide. In 1960, it found, the richest one-fifth of the world's people had incomes averaging 30 times as high as those of the poorest one-fifth. By 1990, the richest one-fifth had incomes 61 times as high. And since then, the gap has continued to widen. In 1998, according to *Forbes* magazine and the UNDP, the three richest individuals in the world had wealth equal to that of the combined GDP of the world's 48 poorest countries.

any mention of the growing disparities of income, in this same economy, between the people who own those global chains and those who eat dinner or sweep up in them.

What makes this view dangerous is that the techno-optimists get their rewards—their income, recognition, power, and perhaps their sense of self-worth—not by manipulating silicon or hydrogen or genes or light to create still more astonishing extensions of our bodily capability, but by manipulating public policies to consolidate support for those new technologies they're confident someone else will achieve. The support they solicit is provided in the form of government subsidies, media hype, and investor confidence. To lubricate that process, however, requires getting rid of any skepticism about whether those new technologies will work. That means not only fending off qualms about whether there will be unanticipated side effects or costs of new inventions; it means dispelling any concern that future solutions to already existing problems may not work—or may be too little, too late.

It's that feverish rush to ever heavier reliance on future intelligence to bail us out of past blunders that creates the techno-optimist market. The faith it has promulgated is so prevalent now that it's hard to see it *is* a faith, and not a system of planning based on fact. It gained currency years ago, after the Apollo space program, with the often-heard comment, "If we can put a man on the moon, we can solve the problems of poverty or pollution." It glosses over the reality that building rockets and building livable communities are two fundamentally different endeavors: the former required uncanny narrow focus; the latter must engage a wholistic view. Building a livable world *isn't* rocket science; it's far more complex than that.

More ominously, today's techno-optimists have narrowed

their focus still further, from a faith in public endeavor (as in John F. Kennedy's "ask what you can do for your country") to a faith in selected private investments. The pursuit of high-tech stock bonanzas in the the late 1990s had more than a little gambling fever in it—it was driven at least partly by the knowledge that over the previous decade or two, astonishing technological breakthroughs had made countless fortunes. In Silicon Valley, over a period of several years, a new millionaire was made every two days. Most investors did not actually envision the next generation of technical power; it was enough to believe that the power was out there, somewhere on the NASDAQ, or the NIKEA, waiting to rocket upward.[4]

In what sense was this faith in technology a form of denial? The short answer is that technology is seen as the key to limitless wealth, as long as it can be used to create new products that everyone will want to buy—and that everyone can buy because the prices are low. Just subsidize consumption, whether of Kansas grain, Saudi oil, Ford cars, or Microsoft browsers, enough to get huge markets hooked. The popular assumption about these prices is that they reflect all the costs of making and getting a product to market, plus profit. (After all, if you're running a business and your costs exceed sales, sooner or later the business fails—so a smart guy like Bill Gates wouldn't give you

4 ❧ About a month before the world's stock markets began wiping out the savings of millions of people, driving thousands of businesses into bankruptcy and putting a number of national economies on the ropes, New York Stock Exchange CEO William Johnston began a commencement speech, "We're not going to talk about a new economy, because there's nothing wrong with the old one. Who would want to change?"

a price that doesn't cover all the costs, would he?) If enough people make that assumption, and sales are good, investors make the same assumption. When profits are piling up, the whole system looks so solid that it seduces us into shutting our eyes to the question of whether there may be hidden costs not reflected in the prices, which someone, sooner or later, will have to pay. When the question did begin to be raised in the late 1990s (as, for example, in the revelation that Silicon Valley has the highest concentration of Superfund toxic pollution sites in the United States), it was quickly quelled by the idea that if we've always been able to solve problems in the past, we'll be able to do so in the future—and the continuous rise of sales and profits that prevailed for half a century after the end of World War II was what proved it.

The followers of this faith are generally more educated, at least in years of schooling, than the religious fundamentalists; they put their faith in technology rather than the Bible or the Koran, but they have much in common with their fundamentalist brethren in one respect: they are insecure in ways that belie their declared assurance that all will turn out well. They are more likely than most to own guns or to live in guard-house communities with private security, to own car alarms and house alarms and other tools of a besieged society; they're more likely to live in enclaves, be suspicious of strangers and hostile to immigrants. They are ready to spend money on things that will let them feel more secure, whether these things be more powerful guns, more tranquilizing drugs, or more reassuring ideas.

Where there's a market of ready buyers, someone will soon appear to sell to them. If it's a market for ideas, the sellers will become organized idea-producers—think tanks or PR consultants that function as the invisible suppliers to mass media. Often, what will appear on the scene is an ideological entrepreneur. When people who had put their faith in technology began encountering unbearable evidence of technology's unanticipated treacheries, the way was prepared for a rescue—the promise of future technology that would more than make up for any imperfections of the past. It was the lure of the double-or-nothing bet for the addicted gambler; no matter how many times you lose, you're convinced you're going to win on the next throw. The ideological entrepreneurs that serve this market—organizations like the Cato Institute on the right, or the World Future Society on the left—know that in this kind of market they can never be proven wrong.

A fictitious world: The largest market for denial is something that, if it were much smaller, might be called a cult of entertainment—a subculture that has swapped the traditional roles of entertainment and nonentertainment. But it's no longer a subculture; in many respects, at least in the United States, it's now the main culture. A few decades ago, we assumed the activities of work, school, companionship, family, etc., were the real substance of our life. Occasional entertainment—the Saturday night movie or dance, the holiday festival, the circus or football game—was an enhancement to add spice to life. Now, for many, the main content of consciousness is sports, games, fashion, movies, TV, "virtual" reality, and vicarious interest in the lives of celebrities—and the rest, including work or school, is only a means of financing the consumption of entertainment. Parents often worry about their children spending too much time watching TV, but in fact the adults watch it even more. In the United States, in the mid-1990s, women over the age of 55 watched 40 hours of TV per week, while men that age watched 35. Younger adults watched 22 to 28 hours.[5]

It's difficult for even the cleverest of info-purveyors to hide real-life threats from us altogether, because while these purveyors control much of the media, they don't control the streets, the climate, or the locks on our doors. Instead of hiding, therefore, one of their most effective techniques—beyond the marginalization and other obfuscations described earlier—is to blur the borders between reality and fantasy, so that in time we are able to react to the "real" with the same detachment we have toward fantasy. It's a process that has crept into news media through such phenomena as re-enacted crimes in which what we see is a splicing of real film and staged incidents, or even a completely staged event presented as "true." As the line blurs and viewers become accustomed to the crossover, the producers can take greater liberties: ambiguities are edited out, action is speeded up, sensational details are given more attention, and sentimental outcomes given more emphasis. The latter, for example, can be seen

5 ❧ The average American child is born into a home where the TV is on 7.5 hours a day. But according to George Gerbner of the Annenberg School of Communications, "the amount of time spent watching TV isn't the main problem. The main problem is that the stories children see and hear are limited to only a few types. . . . Our studies show that casting and fate follow stable patterns, especially in prime time: Men outnumber women two to one. Young people are underrepresented. Older people make up only one-fifth of their actual proportion in the population. Poor people are virtually absent. . . . Television programs—and movies, as well—project the power structure of our society, and, by projecting it, they perpetuate it, make it seem normal, make it seem the only possibility."

in the prevailing coverage of weather disasters like the Red River flood, in which the reporters' focus on the "we shall rebuild" sentiment becomes a subtle form of fictionalizing, in the sense that the reporter chooses which of many possible endings to provide. Even mainstream news becomes "tabloidized."

This has laid the way for a more audacious manipulation in which the very worst kinds of news become fictionalized and therefore nonthreatening. A very real threat, for example, is the spread of dangerous viruses, such as the Ebola. When the movie *Breakout* was released, it enabled viewers to get excited about the threat while knowing subconsciously that they were safe in their theater seats; for all the horror of thinking they were about to see a living person turned to soup inside his skin, they could always walk outside and laugh. Similarly, when the post-apocalyptic film *Water World* depicted a flooded world left by a civilization that had been fatally oblivious to its addiction to fossil fuels, the viewers knew they were only in a theater. Even the mundane excitement of conventional war, urban violence, the breakdown of the family, and the betrayal of the young by their elders who have consumed too much, are all conveyed with this sense of being in a theater and thus not *really* threatened. Movies have always been an escape, but the difference now is that entertainment has become the center of our lives rather than a peripheral amusement. For many, the escape has become the reality.

2. BLIND ACCOUNTING

In our responses to the spikes, we are riddled with contradiction. Most people, so far, have failed to see the spikes for what they are; yet I think most now sense that something enormous is occurring or about to occur. Most of us seem to want to cover our eyes as we accelerate down that curving mountain road; yet we also want to peek between our fingers—to catch glimpses of the approaching switchbacks and cliffs. We do not want to be astonished. The fictionalizing of threats that so dangerously reduces their urgency also lets us, on some level, keep the flame of awareness alive. In that respect, paradoxically, the same infatuation with entertainment that distracts us from real-world concerns may also, on some subconscious level, keep us prepared for the day when we can break out of our reverie. There has to be some reason we're so attracted to epic adventures, even if they aren't yet our own.

As the speed of change accelerated in the last quarter of the twentieth century, the huge anxieties it provided were reflected in this popular culture. One of the early smash hits of the era was the 1981 Steven Spielberg movie *Raiders of the Lost Ark*—the story of a heroic archeologist who seeks a sacred artifact that is also being sought by demonic Nazi looters.

The movie, which became a prototype for most subsequent "action" films, tapped into some deeply affecting myths. Many cultures have stories with the theme that our fate is to endure terrible tests of body and spirit—one after another, tests so horrific they're absurd: the myth of Sisyphus pushing a stone that rolls back on him; the fairy tales about suitors for the hand of the princess put through impossible tasks by the possessive king; the young man setting out to do battle with a dragon. *Raiders* brought them together in a vivid scene—the hero Indiana Jones making his way through a mysterious cave in which he encounters a terrifying arsenal of threats. The scene set a new standard for this kind of storytelling: the roiling snakes, the rolling boulder, the fear of burial, the series of narrow escapes all flew toward the audience with harrowing speed. We felt huge tension, and at the same time we experienced a kind of subterranean thrill of recognition: kind of like *my* life! And of course it was: our lives have always been like that—beating impossible odds.

Consider that more than half the people ever born on earth—perhaps 60 percent—have died young, before reaching childbearing age. If for any one person picked at random (from among all those born throughout history) the odds of reaching childbearing age were less than 1 in 2, then for any 10 people picked at random, the odds of their all having reached that age are just 1 in 9,500. For any random 2 billion of the people who have been born, then, the odds against their all having survived to childbearing age are so astronomically enormous that the number is incomprehensible. Yet, of the approximately 2 billion people who were the parents of today's global population, *every one of them did* reach childbearing age. So, they were not a random selection of people; they were highly selected. We who live now are the survivors of survivors. We, like Indiana Jones, have somehow survived, against all odds, a remarkable series of dangers. Yet, like him, we know—subconsciously, at least—that passing one test leads only to the next even more perilous and more onrushing one.

Another theme making that movie compelling is that the tests are not accidental; they're things the hero could avoid, but doesn't because he's on a quest. It seems to be part of human nature, for those who do not find risktaking unbearable, to go on quests; and perhaps it is beginning to dawn on us that we're on a quest right now that is different from any previous one. It's not the quest of Crusaders or treasure-seekers; this is one on which everyone's survival depends. The tests are commensurate in their magnitude, like inconceivably large boulders hurtling toward us. And, we're not in a theater.

As must now be apparent, the first big test is to identify the real threats to us—a task that has been extremely difficult because of the impediments thrown before us. Even when we have gotten past those impediments, though, there's the huge test of accepting the information for what it is—of not being cut off by the closing steel door of denial. But even when we get past that, the next test faces us: tracking down critical information that is not now available. There, too, we're facing a daunting prospect.

In the global economy, much critical data is never recorded, because of fundamental flaws in our system of monitoring and accounting. Within the fertile but little-known field of environmental economics, a large body of discovery has accumulated over the past few years—most of it simply ignored by the traditional eonomists who dominate government agencies and businesses. The gist of this discovery is that there is a huge, potentially catastrophic blind spot in conventional accounting—a tendency to regard as "free" many basic inputs to the economy that are not free. For example, we think of air as free, because there seems to be an inexhaustible supply of it.

That's what we also once thought of fresh water, arable land, or fur-bearing animals, and it's easy to understand why. Over the past millennium, as human population expanded, the answer to shortages was always simply to expand geographically; there was always more land or water or fish or fur for the taking. When the Europeans took the New World, they thought of it as free. When they wanted fur, they might have had to pay the trappers for their labor, but the trappers didn't have to pay anyone a hunting fee. (And even if they had, as they now do in most places, the fee couldn't possibly pay for restoring a lost species.) The whalers and fishermen didn't have to pay for their use of the oceans.

But it's not just resources that we saw as free; it was also a range of biological, chemical, or physical processes, that would be considered major industries if they were performed by humans: the making of topsoil, the filtering of fresh water to make it fresh; the pollinating of crops. And those perceptions continue to prevail: most policymakers and planners still act as though many of those resources and services were free. After all, if a business doesn't have to pay for them, and therefore doesn't have to make the customer pay for them either, aren't they free by definition?

That's where the blind spot lies. Because the business planner is only concerned with his business costs, he doesn't ask whether the materials or services he's using are really free, or just free to *him*. In fact, what is free to him may be extremely costly to other people—or other species. This has only been acknowledged for a few industries such as cigarette making or the manufacture of ozone-destroying CFCs, which government officials feel they can safely isolate and jettison, like Jonahs, without rocking the boat of free trade. The phenomenon of US states suing cigarette manufacturers for the public costs they incurred as a result of smoking may be only a tiny lifting of the lid on the Pandora's box of unaccounted costs that states or individuals could sue for. Bad as cigarettes may be, the public tempest over them is pathetic in view of the silence that hangs over the enormous damages done by fossil fuels, nuclear power, pesticides, and timber industries.

In the few cases where industries have been called into account for their costs to society, or to the nature that underpins it, the issue has been narrowly defined in terms of specific damages. But there's something more fundamental and pervasive at stake, which is the integrity of the basic economic system as a whole. Focusing on individual industries avoids this larger issue of whether a system that allows business in general to avoid paying its full costs allows far too many vulnerabilities for any system of laws and regulations to rectify the abuses. A system that provides incentives *not* to protect the public is like a boat with a sieve for a hull, in which the only way to avoid sinking is to plug all the holes individually and then try to keep them plugged. Better to build the hull with a watertight material in the first place. In economic terms, better to start out with full accounting for all industries as a matter of course. That means tallying known costs of a product and requiring the industry to include those costs in its prices to consumers up front.

The higher price then becomes a big incentive to reduce the costs that are external to the industry and hidden to the public. How big? One way to get a sense of that is to look at how much of our consumption is falsely regarded as free.

In 1997, economist Robert Costanza of the Institute for Ecological Economics at the University of Maryland released a landmark report that estimated, for the first time, the dollar value of the services performed by natural processes worldwide. Synthesizing the findings of more than 100 studies, Costanza and his team found that the value of services performed by natural processes, and mostly regarded as free, actually exceeds the total of all human industry. It was estimated to be in the vicinity of \$33 trillion per year, as compared with \$25 trillion for the global GNP (or GWP). This finding is staggering to conventional economics. Yet it may be a measure of how far our hubris has gone that we are surprised to find the value of Nature's works rivals that of our own.[6] And this value doesn't include the full value of the resources themselves, but only of the processing they provide.

6 ❧ It may also be arrogant to put a dollar value on nature at all—first because it will always be an undervaluation, since we'll never know all there is to know about what it does to make the natural world what it is, and second because dollar values imply that all debts can be paid. No value can be assigned to "restoration" of a species that has been extinguished.

In other words, it doesn't include the value of natural capital, which really can't be measured. And our industries are very busy spending that capital.

Ecological economists say that if full accounting were instituted, a sea change would occur in human behavior, at least to the extent that behavior is influenced by economic incentives. Most of what we consider "bad" (such as polluting groundwater, or destroying habitat by bulldozing forests for new development) would become much more expensive, and there would be a sharp drop in our incentives to do it. "Good" behavior (such as generating electricity without emitting CO_2) would be relatively much cheaper, so the incentive to be good would be higher. David Malin Roodman, in his book *The Natural Wealth of Nations*, notes that in the current economy, many costs of industrial activity are not incorporated into markets—a fact that may help to explain why market-watchers have been so bewildered by such events as the collapses of global stock exchanges that swept the world in 1998. "By translating environmental costs into prices, governments can grab decision-makers by the bottom line and help consumers better understand the true environmental costs of their purchases and investments," Roodman argues. "If we're going to save the planet, prices must tell the ecological truth."

This sea change, which would bring into view a vast category of critical information now either ignored or hidden, could be particularly critical in the behaviors affecting the four spikes. Here is what might occur:

- Burning oil and coal imposes not only the unknowable future costs of climate change, but the well-known present costs of pollution—such as the reduced lung capacity of children who live in Los Angeles or Mexico City, or the increased risk of heart diseases among their parents. If we add these fuels' share of the world's health bill, and then ask the fuel producers to pay that bill and pass on the cost to their customers, we could expect the price of gasoline, coal-generated electricity, and heating oil to rise sharply. That would make wind and solar energy relatively cheaper (they're already closing in, even without the aid of a level playing field). It would make hybrid electric cars immediately more attractive. Investment would pour into the clean energy alternatives and out of the obsolescent ones. Ripple effects throughout the motor vehicle, utility, and construction industries, among others, would become an economic tsunami. The price shift could be pump-primed by a tax shift (from earning income to burning oil, for example). The trigger force is self-interest, but the impact is far-reaching, because it provides the basis for the major reduction of CO_2 necessarily to stabilize climate change.

- Supporting a meat-heavy diet imposes a wide range of costs not included in the prices consumers pay in the store or restaurant. Some of those costs return years later in the form of hospital bills or lost income resulting from heart attacks, strokes, and cancers. These costs are borne mainly by the heavy meat consumers and their families. But as meat consumption grows worldwide (it has increased five-fold since 1950), other costs are imposed universally. Because meat eaters take much more farmland to support than do their vegetarian counterparts, and because we face severe shortages of grainland as population grows, the demand for meat-rich diets increases incentives to clear still more of the world's forests and wetlands for farming. (Worldwide, more deforestation is done by agriculture than by the timber industry.) That, in turn, undermines the "free" services performed by forest systems, including filtration of fresh water and provision of habitat for wild pollinators that service a majority of human food crops. If we add these costs to the price of meat sales (with the revenue directed back to public health and land conservation), meat consumption will fall. Public health will improve markedly in the high-meat-consumption countries; millions of people will become leaner and more vigorous; many families will enjoy the vital presence of elder members otherwise lost to untimely death or infirmity; pollution of rivers and seas by livestock waste will be reduced (thereby alleviating marine diseases like pfisteria and enabling many fish populations to rebound); and the widening of the gap between Earth's population and its carrying capacity will be slowed.

- Often, we try to stem overconsumption with well-meaning gestures that are then overwhelmed by still more consumption. US laws requiring higher fuel economy in cars, for example, produced significant energy savings in the 1980s. But those savings were wiped out in the late 1990s when people switched from small cars to big sport-utility vehicles (SUVs) and pickup trucks. Suppose, however, that instead of getting into that kind of Sisyphus-like position, we raise the price of overconsumption across the board. Excess consumption is not so simply defined, but that needn't keep us from raising prices to pay for its uncounted costs. After a moderate allowance, make the price rise steeply. What's moderate? It depends on what's

> sustainable. Some would suggest that if science tells us we can produce 100 million tons of paper and paperboard (cardboard) per year worldwide, without any further degradation of natural forests, and with an adequate allowance for returning some degraded or plantation area to natural growth, then that's how much we should allow into the market globally at "competitive" prices. Beyond that, a steep consumption tax, and maybe added taxes on production and advertising as well, could be imposed. Others argue that we shouldn't trust the market to do this, but should set absolute limits on "throughput"—and then let the prices find what level they will.

Actually attempting to implement any of these shifts, in the present information environment, would be difficult and maybe futile. (In the United States, for example, proposals for energy taxes have been quickly shot down, and consumption taxes haven't fared much better.) The point, right now, isn't to propose policies but to show the logic that will have to underlie any workable policies—that if the costs of something aren't visible or felt, they tend to rise unchecked. Only the visible part—the part paid by the immediate consumer—stays low. If the information climate is changed to make the costs of excess consumption visible, we might begin to see cultural attitudes change in turn—and what seems politically difficult would then would become politically supported.

Not only do conventional measures of accountability overlook huge costs, however; they also treat accountability with a built-in cynicism by treating all productivity as morally equivalent. A dollar is a dollar, or a ruble is a ruble, whether produced by a life-saving drug or a killer one. Here, too, large costs are hidden. However, here the problem is not just that costs may be imposed on other people or species who are not parties to the deal. It's also that no distinction is made between costs to one person that are benefits to others (such as the costs of food when paid to a farmer who is using sustainable practices), and those that become a loss or hurt to someone else (such as the costs of food produced by industrial monoculture practices, when paid to producers whose business profits from driving small family farms out of business).

This makes for an accounting task even tougher than that of uncovering hidden costs, because it means pricing not only to cover the damages of assets formerly considered free, but also to distinguish between recognized financial costs that make net additions to public wellbeing and those that undermine it. This is pricing with a moral component, and what it says is that even if we accept the idea of a free market economy, it should never allow one person to get rich by means that make other people, generations, or species get poorer. There are few economists in the world today, even among the most myopic ones, who don't agree that wealth *can* be built (or rebuilt, if we look at what has really happened) in absolute terms. The key is to de-link the building of wealth from the spending of natural capital.

A third area in which accountability has failed us is in our monitoring of critical global indicators. In other words, not only our means of measuring human economic activity are faulty, but our means of observing them so they can be measured. It's already clear, from the look at GDP, that accountability is not just a matter of watching the money, but of watching the human behavior that produces it.

For each of the spikes, there are key indicators for which current information gathering is dangerously leaky—and for which powerful global surveillance is essential:

Carbon Gas: Who's keeping track of emissions? The climate treaty negotiations were organized around the idea of national emissions quotas, which implies that those national outputs can be known and verified. But can they really? Certainly not from the ground. Consider China, which now has a larger population than the entire world had at the onset of the Industrial Revolution, and which is in the full throes of its own nineteenth century-style industrial revolution. China plans to build 500 new coal-burning power plants to add to the thousands of plants and millions of industrial and residential chimneys and smokestacks already in place. Recall the farcical attempts of a UN inspection team to monitor a handful of sites in Iraq, when they first suspected that the country's military was building nuclear, chemical, or biological weapons. (Recall too the failure of the CIA to see signs of India's impending nuclear tests in 1998, despite the fact that the agency knew well the technological capability of the regime and the belligerent mood of the new government.) The number of people it would take to monitor China's carbon gas emissions, a form of chemical release far more potentially destructive than anything Iraq might muster, would be like that of an invading army—and likely just as unwelcome. Monitoring devices can be disabled. But even if China agreed to let such eco-police in, the effort would likely be futile. Coal smoke may be black, but the black part is soot. A power plant could be completely cleaned of soot and still emit large amounts of invisible CO_2. That is what has happened with American cars.

Then there's the puzzle of carbon sinks. The prospect of trying to keep track of the carbon stored in sinks became a true accounting nightmare-in-the-making during the climate treaty negotiations of 1997 and 1998, when some countries began to warm to the idea of using sinks as "credits" toward their assigned levels of greenhouse gas reductions. With the treaty requiring a 5-percent reduction from industrial countries' 1990 levels by 2010, negotiators hit upon a clever accounting precedure that could conceivably save them the trouble of asking their industries to reduce emissions at all. The idea was that when it is time to take stock of progress made, in 2010, it will be permissible to count the carbon absorbed by tree planting. But in establishing a 1990 level to be compared against, the carbon stored in trees would *not* be counted. For heavily industrialized countries, where tree plantations are a common source of pulp for paper manufacturing, this "gross-net" accounting does just what one of those "before-and-after" weight-loss ads does, when only in the "before" photo is the subject wearing bulky clothes. In the "gross-net" system, only in the "before" assessment (1990 levels of emissions) will the carbon in trees be counted—thereby creating an automatic proof of "progress" made in the country by 2010, when that carbon is not counted and therefore is subtracted from the totals. In the one industrialized country (Australia) where there's not enough tree planting for this fraud to be useful, it was agreed to allow "net-net" accounting. In effect, each country was able to choose whichever accounting system would allow it to show progress toward its target—by either counting or *not* counting changes in the amount of carbon stored in trees—without actually reducing emissions from factories or motor vehicles. And, since carbon sinks also include peat bogs, soil, and ocean water, it's clear that any serious attempt to track the global carbon flow on a nation-by-nation basis—aside from its inherent Rob-Peter-to-Pay-Paul absurdity—would likely be futile. That's not to say there isn't a compelling need to regenerate natural forests, as opposed to monoculture plantations. But planting millions of trees can never address the problem of rising CO_2 *accumulation* (the addition of carbon not already in the planet's active natural cycle of plant growth and decay), which comes from exhuming and burning the fossil trees of thousands of previous centuries.

As difficult as the monitoring of CO_2 will be the observation of its immediate effects on climate, the cyling of fresh water, and photosynthesis. Current capabilities to track forest fires, hurricanes, tornadoes, and floods are still hit-and-miss, but will need to be made seamless—along with seamless universal warning procedures. Otherwise, with the combination of population expansion and increasing storm intensity, the frequency of incidents in which populated areas are taken by surprise—like the time a swarm of tornadoes and floods swept northern India and snuffed out 10,000 people in a night—are likely to rise sharply. Global surveillance will also be important in tracking changes that seem more gradual from the perspective of human perception, but which in geophysical time are shockingly sudden: the spreading edges of deserts, deforestation, shrinking lakes, fish-killing algal blooms, coral bleaching, growing dead spots in seas and lakes, and infestations.

Population: Don't think of the population spike as an abstraction, but as a pattern of regional upheavals. Yes, on our graph it appears as a singular global phenomenon, as though humanity were spreading the way a cancer spreads through a body or an algal bloom spreads over a lake. In one sense, the population spike is like them: all involve surges of unsustainable growth that, if not brought under control, ultimately consume their host.

But what the spike really is, on the ground, is the sum of a pattern of hundreds of separate breakdowns in the relationships between regional populations and their environments. As in the breakdown of a marriage, there may be simply a gradual deadening or there may be a nasty, even violent, falling out. It can result from "too much take, not enough give," as when the population gets too big to be supported by its local crops and begins selling off its natural capital in the form of subsidized exports to pay for more imported commodities. Or it can result from a form of cultural philandering, in which a few members of the local population conspire with outsiders to exploit local labor or resources in ways that enrich those few while impoverishing the rest. Or, the breakdown can be violent and traumatic, resulting from resource scarcities, diseases, or disruptions of climate caused by the cumulative effects of population imbalances in all the other regions. As a result of these regional breakdowns, the relationships between regions, too, are in continuous flux. The people move across borders in waves of migration or flight, and their environments move in patterns of wind and water currents, climate-driven contractions or migrations of ecosystems, and bioinvasions.

Both kinds of movement have increased, and both cross national boundaries with impunity. To track these movements requires on-the-ground monitoring within nations, but more importantly it requires an integrated global system that can follow cross-border movements without getting bogged down in the red tape of 180 different governments. Changes of population, whether due to movements or reproduction, will have to be coordinated with the ebbs and flows of resources, weather, and pollutants, in much the same way that the hundreds of airplanes going in and out of a major airport need to be coordinated by a single air-traffic control system. To continue allowing these flows to be monitored and controlled mainly through bureaucratic transactions between individual nations would be as hopeless as asking individual aircraft to negotiate with each other as they approach an airport, over who will land on what runway in which direction and when. The result would be chaos.

Extinctions: Watching this spike is like watching the simmering of a large cauldron of stew and trying to keep track of the individual ingredients and what's happening to them. At the very least, it will require monitoring a number of interdependent phenomena. One is the shrinking of forest cover, which is critical not only to habitat but to sequestering carbon gas. (The more the planet is denuded, the less carbon is stored in trees and must be loosed into the air—it can happen when the forests burn and aren't regrown, but also when the organic matter in a forest floor dries up and releases its carbon after the tree cover has been cut, or when wood or paper decays in a landfill.)

A second indicator is the fragmentation of what forest cover remains; as habitat is crisscrossed by roads and development, and turned into "terrestrial islands," it becomes far more vulnerable. In *World Watch*, I referred to this once as "the fragging of the Earth"—using a term that came into American usage for a few years during and after the Vietnam War: "it referred to the notorious practice by which soldiers—whether in anger, disillusionment, or war-stressed uprooting of cultural values—sometimes tossed hand grenades at the feet of their own officers. On a larger landscape, the fragmentation of life continues. As more space is carved out for new streets, housing developments, and other 'development,' forests are cut apart—bulldozed into fragments that may look pleasant as backdrops to suburban living, but have been ecologically exploded. Because many species cannot survive in habitats reduced below a certain minimum size, the thousands of small fragments of woodland that are scattered through developed areas can no longer harbor the biological diversity they once had. They may be superficially attractive, but beneath their green boughs they have become irreversibly weakened." Satellite photography can provide highly detailed mapping of both fragmentation and overall loss, providing information that will be essential to developing regional land-use plans.

A third indicator that will require close watching is the slow but inexorable migration of climatic zones. As the planet warms, the zone in which a particular species finds a habitable temperature range moves farther from the equator. If that species can move as fast as the temperature shift, those of its population that find the southern border moving out from under them may be able to survive by migrating north (in the northern hemisphere). But if the species is too slow, as many species of plants may be, whole communities could get cut off. At a warming rate of half a degree per decade in the middle latitudes, for example, vegetation zones would be displaced by anywhere from 90 to 340 miles in a century. But many North American tree species moved only about 20 miles in a century when the last Ice Age ended. Some long-lived tree species can migrate only about 3 miles in a century. John Ryan, director of research at Northwest Environment Watch in Seattle, notes that such trees as beech and oak, which can migrate only the distance their seeds can be scattered in a generation, would be unable to move quickly enough and would suffer severe die-offs.

The same is true of retreats to higher altitudes, and we've already seen evidence that as ecosystems retreat uphill, some of their members will not make it. For some species, the loss may be unnoticed. But the monitoring will directly impact on humans in at least two major respects: it will affect where we can grow crops, and it could have major impacts on the ranges of pests and diseases. Again, the patterns of these migrations are irrelevant to national borders.

Still another category of key indicators for extinctions is a kind of disruption for which we several years ago coined the term "bioinvasions." It's a kind of disruption that's as old as life, but in recent years it has accelerated hugely—and constitutes a major reason why extinctions have accelerated. A bioinvasion occurs when a species is transported by some means from its own native ecosystem into one to which it is alien—for example, a fish that is native to a river in Africa gets tossed into a canal in Florida. It can range from the transport of infectious microbes in human blood to the spread of snakes in cargo. In most cases, nothing remarkable happens: the alien organism is quickly wiped out by antibodies or predators to which it has developed no defenses, or it can't find its accustomed food, or it is too warm or too cold. But occasionally, the reverse occurs—the invader thrives and finds a place that has no defenses against it. The resulting proliferation can be a nuisance (starlings, honeysuckle), but occasionally it becomes a disaster. Bioinvasions can turn into outbreaks of crop-consuming insects, killer diseases, or destroyers of entire ecosystems. Chris Bright, in his book *Life Out of Bounds: Bioinvasions in a Borderless World*, notes that the introduction of a new species can be like a bomb. "Most are duds, but one in a hundred goes off." The problem is that so many leaks of one ecosystem into another now take place that one in a hundred adds up to a devastating number of "live" bombs.

In centuries past, ecosystems were kept separate by natural barriers (oceans, mountains, elevations, temperature zones), and only rarely was a species successfully carried over these barriers—clinging to a log floating across an ocean, for example. Today, organisms are carried in huge quantities by the mechanisms of global trade, tourism, smuggling, and human migration. Because modern transport is quick enough to carry viruses, insects, and rats around the world in a few hours, before they die, the geographical isolation no longer is effective. Every country in the world is more vulnerable now than it was a few decades ago. Where defense systems are peering at the skies to defend against ICBMs, they really need to peer more closely at what's in the blood or boots of airline passengers, or in the ballast water of ships. Far more thorough monitoring of traffic flowing across natural borders is needed—not to screen passports, but to catch biological hitchhikers.

3. THE GLOBAL SHADOW ECONOMY

Finally, in addition to identifying critical phenomena, recognizing them for what they are, and shoring up our monitoring of data where it is now porous, there's the problem of information that has disappeared altogether.

It may run completely contrary to our impressions of the world to suggest that in a time of instant, global communications that are unimpeded by either natural barriers or national borders, a huge portion of the information that is out there now—including information vital to public wellbeing—is sequestered. In some ways, the situation has actually been worsened by the proliferation of satellites and the advent of the Internet, because we now have access to so many sources through our computers that we may be inclined to think nothing is any longer hidden from us. That impression is nicely reinforced by all the stories we've heard about e-mail being used by people in repressive regimes to communicate with the free world right under the noses of their governments. And, there are stories like that of the "most-wanted" murderer who fled the United States to a remote village in Guatemala—only to be quickly apprehended after a villager saw the fugitive's picture on the Internet. These, along with all the alarming reports we've read about how even your most private financial or medical information can now be accessed ("The Death of Privacy," warned a 1997 cover story in *Time* magazine), convinces us that nothing is secret anymore.

But in fact, secrecy has grown on this planet like a tumor. The sheer volume of the information we do have flying at us (whether on the Internet or TV or on a drive through any strip mall) very successfully distracts us from what we're not seeing. Moreover, even a lot of the Internet information we're accustomed to having access to may begin to recede from sight, as encryption gains wider use—presumably to protect privacy, but perhaps to keep out government regulators as well.

If the sequestering of information were random, that might be something to shrug off: even what's readily available may be astronomically more than anyone could absorb in a lifetime. But what's hidden is not random. Much of it is information that—whether or not this is the reason for its being hidden—is critical to our ability to deal with the spikes.

The explosion of information is a measure of the same explosions of human activity that drive the spikes. But much of that activity is taking place beyond the grasp of accountable public institutions: it is untaxed, unregulated, and—often—unseen. Some of it is illegal; some is simply "informal"—the activity of the hundreds of millions of people in the world who are not a part of the escalating global economy. They are the people who do not have tax-ID numbers, credit cards, or e-mail addresses—or sometimes any addresses. Some sociologists say it's wrong to lump together the struggles of impoverished street vendors or subsistence farmers, or the world's remaining hunter-gatherer populations, with the activities of drug cartels or illicit weapons traffickers under the broad umbrella of a "shadow economy." But as globalization continues to gain momentum, some of the same conditions that cause hundreds of millions of people to become economically marginalized also allow millions of others to become deliberately evasive. Information that is kept in the shadow, whether by the negligence or ineptness of governments or by deliberate evasion, can be critical to stabilizing the spikes:

Population: Studies in the 1980s and 1990s found that the explosion of squatter settlements accounted for 70 to 95 percent of all new inhabitants in the cities of the developing world. And in many cities, 30 to 60 percent of the whole population now reside in areas outside the reach of any public authority. One consequence is that large numbers of people are not getting the services needed to stabilize family size—the access to family planning information, contraceptives, and educational opportunities that liberate girls to pursue roles other than those that trap them in repeated and closely spaced pregnancies. Another consequence: squatter populations are unstable, because their settlements are not ecologically based (they are not fishing villages or farming communities or in any way rooted to the place where they've washed up), and that makes them more likely to migrate toward more affluent places. If the migrations were trickles, they might be absorbed. But in the numbers now imminent, they can overwhelm places that were stable, accelerating breakdown in the sense of rootedness that is essential to our collective survival. In short, while overpopulation is inherently dangerous, overpopulation that is in chaotic motion is more so—both because it grows faster and because it heightens the risks of synergistic unraveling of once-stable systems.

Carbon gases: In the 1990s, the burning of forests grew enormously, much of it as a result of slash-and-burn incursions by outlaw farmers into the remaining virgin rainforests.[7] In 1995, satellite photos revealed that these settlers—the advancing edge of humanity's final spasms of expansion—had started 70,000 forest fires in the Amazon alone. The fires sent up a shroud of carbon-laden smoke large enough to blanket 2.8 million square miles of the planet's surface. The following year, the number of fires increased. In 1997, the amount of burning in the Amazon increased by another 28 percent, and the amount of destruction caused by wildfire worldwide was the greatest in history. Then, in 1998, came the vast fires that swept much of Indonesia and Malaysia, in the aftermath of which the Indonesian economy collapsed and the Suharto government fell. In the Southeast Asian fires, like the Amazonian ones, most of the fires were deliberately set, in this case not by poor farmers but by rich plantation-owners intent on expanding their tracts. But again, the activity was illegal. Both among impoverished people and among corporations, shadow-economic activity is on the increase, and governments seem helpless to stop it.

Consumption: The essence of excess consumption is that it is unrestrained either by disincentives established by society or by the consumers' own sense of responsibility. Societal disincentives—such as full-cost pricing—even when enacted, may not reach

7 ❧ Unlike the periodic burning of temperate forests, which has been described by ecologists as a natural process that helps the ecosystem maintain its long-term vitality (as was judged to be the case in the 1989 Yellowstone National Park fires), the burning of a tropical forest is not natural, and it is fatally destructive to the tropical ecosystem. It is global warming-induced drought that has allowed forests in Indonesia and elsewhere to dry enough to burn, and most of the fires destroying tropical forests in the 1990s were deliberately set.

people in the informal or illegal sectors. For example, if wood is priced to reflect its full costs, the legal timber industry may slow its cutting of forests. But Third World settlers who burn trees to clear land aren't affected by that—and in their neck of the woods, the deforestation will continue unabated. It may even *increase*, as the higher prices of wood open up a lucrative market for black-market timber. For disincentives like full-cost pricing to work, they will have to be universal, so that they can't be undermined by smugglers.

Extinctions: The same spread of frontier populations that produces fires and deforestation also, by eliminating habitat, hastens destruction of the planet's richest stores of biological wealth. But the shadow economy threatens species in other large ways as well. Illegal wildlife trade, in particular, has become as much out-of-control on a global scale as the looting of stores sometimes is in an urban riot: it is blatant and frantic. One reason: the prices for rare species—coveted by private collectors—now often exceed even those for narcotic drugs. A small smuggled arawana (an endangered Asian tropical fish), for example, can bring up to $10,000 from a US buyer. The body parts from a tiger can bring $1 million. The problem is not just that species are kidnapped from their native habitats, and locally wiped out, but that when they are taken into unaccustomed habitats they may destabilize their new environments by attacking species that did not evolve with them and have no defenses. The traffic in biological products is rising like a tidal wave, and the more mixing takes place, the fewer species survive overall.

Of course, there has always been a shadow economy. But in the modern world, the risks are larger because the scale is larger, and because the damage that can be done by unmonitored groups—thanks to their access to high-powered technology—is vastly higher. In this respect, there's a blurring between shadow activity and privatization. Many activities that a government might consider illegal could be characterized by a corporate lawyer as merely "proprietary." And increasing amounts of activity, because of the rising pace and volume of change, haven't been addressed by governments at all: they are neither legal nor illegal; they fall deep between the cracks. As global commerce is increasingly run by global organizations that are not controlled by any single government, it sometimes seems that more falls between the cracks—and out of sight—than doesn't. The bottom line is that if governments can't keep track of what's happening to large flows of organisms and substances that need to be stabilized if the planet is to be stabilized, their ability to develop effective mechanisms for stabilization will be treacherously compromised.

If there's a mystery as to why shadow activity is expanding fast, the explanation may lie in the fact that two sea changes are happening in tandem, with a powerfully synergistic effect: As borders become more permeable, the outcome toward which they are moving is not a borderless world as some futurists have envisioned, but a world in which the nature of borders has changed profoundly. The borders of the future will not be rigid structures like the Great Wall of China, the Berlin Wall, or the hapless barriers on the US-Mexican border; they will be permeable and fluid, like the membranes of living organisms—operating by the laws of nature rather than of man. And as that shift occurs, the nature of human loyalties is changing as well.

7

KEEPING UP WITH SHIFTING POWER

To SUMMARIZE so far: we are encountering world-changing phenomena of a magnitude never seen before—phenomena that should astonish us yet do not. They are being kept out of our awareness by the marginalization and fragmentation of important information, by our declining ability to keep up with the speed of change, by our lapses into fantasy and denial, by difficulties in tracking critical data, and by the disappearance of critical data. The extraordinary kinds of remedies needed are dictated by the extraordinary nature of the problems: we need to shift the whole thrust of our information media, the priorities for research and monitoring, the mindsets and budgets of our military and security agencies, and what we mean by education.

Yet all along, we may be assuming that the institutional framework within which these shifts take place will remain basically the same. It may be as hard to imagine today's institutional framework changing radically as it is to imagine today's technological capabilities not changing radically. We've gotten used to technology changing at mercurial speed, while institutions remain hidebound. But many signs now indicate that the difference may only turn out to have been a lag. After all, if technologies extend the capacities of individual humans, whereas institutions extend the capacities of whole populations, then the development of a technology is by far the simpler task. So, it makes sense that institutional innovation would come later, after the technological evolution has had a chance to shake out out some of its failures. Soon the institutions of society may begin to change as radically as technology has and that could lead to a whole new set of pitfalls. The path out of the cave, into the sunlight, leads to the brink of a chasm.

We now have good reason to believe that even among agencies that are confronting climate change and biodiversity loss, most of the planning is focusing on problems that may not be the main problems created by those phenomena—that, rather, the biggest problems may turn out to be shocking surprises. But even if the phenomenon is recognized and the problem is correctly identified, time and energy may be wasted because of questionable assumptions about the institutional mechanisms that will be used to solve the problem. It's hard enough to motivate leaders to work on the right issues; it's harder still when the institutions they lead are moving on shifting sands.

In the United States, for example, policymakers spend great energy researching and debating such questions as whether the Social Security Trust Fund will run dry in 2020 or 2025. A small number of people ask whether there will be a US Social Security system then. Almost none ask whether there will be a viable United States—or if there is, whether if will operate under the same constitution. Anyone who asks such a question risks ridicule. Yet, only a couple of decades ago, when Soviet grain output had reached 210 million tons per year (up from 100 million two decades before that), planners were debating whether their output by the end of the century would reach 290 million or 300 million tons. None dared ask whether there would be a Soviet Union, or foresaw that by 1996, total production from all the land that had been Soviet would have fallen to a famine-inviting 120 million tons. It may have been the right debate, but it was the wrong institutional body. That body no longer is.

It would be presumptuous and naïve to think that we can now foresee events like the Soviet Union's fall, when we couldn't before. If anything, the faster pace of change now may make it harder to see than ever. But we're also higher on the learning curve in understanding the nature of change, if not in predicting the specifics. We know that while the biggest changes can be surprises like the Soviet collapse, the forces that cause those surprises are not out of the blue at all. Once we learn to see through the obfuscation, we can detect clear patterns of change, and if we revisit those past surprises, we can see that even the surprises fit the patterns. The fall of the Soviet empire, for example, may or may not have been a victory of Free World ideological warfare, or of the manifest destiny of global capitalism. But it was clearly consistent with a pattern of declining sovereignty of national governments everywhere. Even as the United States basked in its ideological triumph, it found itself with less power to control the course of world events than before. This is not to say that the United States will follow the Soviets into sudden collapse, but it's a fair warning that the most important changes that will occur will be of a magnitude completely outside the range of possibilities within which planners concentrate nearly all of their attention.

The collapse of any one nation will inevitably come as a surprise to many observers, since collapse usually is preceded by a certain degree of obliviousness. But that more nations will collapse in the next decade now appears highly probable. As noted earlier, we take it for granted that the nation is the fundamental unit of human governance. But memories are short, and we forget that nations as we know them today are a fairly recent institution. In the past few millennia they have succeeded city-states, tribal lands, empires, feudal kingdoms, forest commons, and nomadic no-man's lands. The nation, too, will have successors—and even if that doesn't happen all at once, the process of eroding national sovereignty could profoundly alter our prospects for dealing with the unraveling of human security.

That erosion is already under way, as nations are yielding power to other kinds of governance:

- *International organizations*, including the World Trade Organization (WTO), World Health Organization (WHO), World Bank, International Monetary Fund (IMF), and the United Nations (UN). While these organizations began as networks of nations, they have become more autonomous, and individual nations have become less able to regard their services as optional. Indonesia has depended heavily on IMF bailout loans. Russia has had a lot of help from the World Bank. US laws have sometimes had to yield to WTO rules. As nations undergo a partial shift from client to supplicant, international organizations are looking more and more like true global organizations.
- *Culturally distinct localities* that in recent years have become increasingly defiant of, or disconnected from, the nations that rule them—the people of Chiapas in Mexico, Chechnya in Russia, Ogoniland in Nigeria, East Timor in Indonesia, and thousands more that form an invisible, underlying map of humanity that is more substantial than

the map of nations. There is a revealing story from the Amazon. A few years ago, after Venezuela and Brazil agreed to grant more autonomy to the indigenous Yanomami people whose territory spans their borders, a reporter trekked into the Venezuelan Amazon with an interpreter. He found a Yanomami and asked, "Did you hear the news? Venezuela has given you independence!" The Yanomami, drawing a blank, replied: "What's Venezuela?"

- *Global corporations*, many of which rival nations in overall economic and political clout. Ford Motor Company, for example, now operates in 100 countries. Your Visa card is good in 224. But such numbers don't just measure the reach of a US- or Japanese-based company's sales arm. The trend in recent years has been toward a globalization of everything from executive management to parts procurement to assembly, so that the idea of a corporation having a home country is becoming rapidly obsolescent. David Korten, in his book *When Corporations Rule the World*, observes that this trend toward "transnationalism" means that companies like the giants of the Business Roundtable or the Global Climate Coalition are becoming increasingly evasive about obligations to the nations of which they were once citizens: "Although a transnational corporation may choose to claim local citizenship when that posture suits its purpose, local commitments are temporary, and it actively attempts to eliminate considerations of nationality in its effort to maximize the economies that centralized global procurement makes possible." Korten further notes that the world's 500 largest corporations employ only 0.05 of 1 percent of the world's population, but control 25 percent of its economic output.

- *Regional trading markets,* including the European Union (EU), Asia Pacific Economic Cooperation Forum (APEC), and North American Free Trade Agreement (NAFTA). Nominally, these amorphous organizations are coalitions of national governments. But on closer examination, they work mainly to the advantage of transnational corporations, while bypassing many of the interests of local economies that national governments are presumed to protect—and also bypassing the accountability of government. APEC, for example, is both highly secretive and quite nonbureaucratic; it has been called an "ad-hocracy" because its decisions are made by ad-hoc groups of top government and business leaders (rather than permanent departments or agencies) that can dissolve and disperse when finished with business—leaving no record of what they have agreed to, and no obligation to explain it.
- *Nongovernmental organizations,* which in financial clout are poor cousins of nations and corporations, but which have exercised rising moral force and capacity to influence policy: the International Union for Conservation of Nature (IUCN), Worldwide Fund for Nature (WWF), Amnesty International, Greenpeace International, World Council of Churches.

Political scientists have had their eye on this flux, and their studies have chronicled the declining sovereignty of nations. The general outlook these studies convey is that we're undergoing an institutional shakeout of which such developments as the breakup of the Soviet Union, the rise of the EU, and the growing virulence of international terrorist movements and rogue nations are only the opening volleys. As the shakeout progresses, much of the power once exercised by autonomous nations will shift to other political entities. That could have large consequences for plans like the global climate treaty, which is only an agreement between nations, not among the corporations or cultures. So far, such ramifications seem to have had relatively little sway on the officials in national governments, who continue to behave as though their agencies will be administering in 2020 more or less the same functions they do now.

Yet, even a full awareness of the shifts in institutional organization now underway could fail to register the extent to which the overall governance of human economic and social behavior is likely to shift, because many of the functions once performed by political institutions at all levels are metamorphizing. Global media, commercial advertising, and the culture of entertainment are rapidly supplanting traditional institutions of authority from the family unit to the community to the nation. Families, cities, and nations all have established methods of law enforcement and punishment, for example, but some of their newly ascendant surrogates do not. This raises perplexing questions about who will assume responsibility for the consequences of human behavior.

Think, for example, of the likely consequences of behavior—not defiant or criminal behavior, but simply normal day-to-day activity—of someone whose most vivid or engaging experience takes place in a world of fictional drama, staged conflict, manufactured sentiment, and other artificial excitements. For this person, a life of TV, cyber-surfing, stadium concerts, theme parks, and shopping malls has almost no visible connections to the physical and biological world that supports it. The older institutions of family, community, and state did, and had rules and laws—however badly conceived they may have been—to keep individual behavior harnessed to the family's or society's common needs. The harnessing was often a shackeling, and sometimes amounted to a massive abuse of a kind that continues today in the marginalization of a billion subsistence farmers. But abuses aside, the connecting of people's activities to the wellbeing of their communities was a vital function. It was important for children to grow up knowing that they must not pee in the well, or set fire to the barn, or pat the baby bear, or forget to milk the cow. They knew that there are consequences for things done badly and for things done well.

In a world where TV and PR have replaced parents or elders, and advertising attracts vastly more attention than government, there are often no clear consequences. For better or worse, the traditional institutions have lost their grip. That may be good news for many individuals, if it means impoverished dirt farmers can drop their hoes and head for the better life of the city as has been happening in China. It could be bad if it means the teenagers driving their 200-horsepower SUVs to the mall haven't the slightest idea what they are doing to the world's hydrological cycle and biological health—and to their own possibilities for connection to real people and life.

Beyond these shifts, some observers have seen seminal changes in human "wiring"—in consciousness itself. The New Age movement, the rise of religious fundamentalism, and the burgeoning of cyberspace all in different ways increase the likelihood that the minds of too many have drifted too far from any common ground with traditional government to be able ever to fully connect again. What that could mean is that the planning now being done, aside from being too marginalized and gutted, is too geared to the mechanisms of specific agencies that may have changed beyond recognition—or even disappeared altogether—in another few years. Global security demands that planning move to a level that transcends the political fates of particular governments or agencies.

To prepare for the information management of the future, then, it is essential to anticipate the changing structure of governance in the twenty-first century. For example, consider how that changing structure could affect the flow of information about fresh water resources, intelligent management of which will be absolutely critical to our ability to ride out the spikes long enough to bring them under control.

We think of water as "belonging," like land, to particular territories: Minnesota and Sweden have lots of lakes; Southern California and Mauritania have few. But that's a fallacy, like saying your left leg "has" two pints of blood. If the blood didn't circulate through the rest of your body, the leg would die—and so might you. The water in Sweden has to circulate through the planet's hydrological cycle just as surely, or large patches of the planet will die. It's as risky for nations to own water in a way that severely disrupts its flow as it would be for a leg to prevent blood from returning to the heart.

On the most visible level, the problem is simply that major rivers flow across state or national borders, and heavy withdrawals by the upstream jurisdiction can mean shortages downstream. If Ethiopia takes too much from the Nile, Egypt is deprived. If Las Vagas takes too much from the Colorado River, the

Colorado Delta in Mexico is deprived. And, in fact, it is deprived—it is now as dry as desert. Yet, nearly 2.4 billion people—as many as would populate 9,000 cities the size of Las Vegas—now live in river basins shared by two or more countries.

But the problem is bigger than just an inequitable distribution of water to different users, devastating as that is. Water shortages have a ratcheting effect on the impacts of each of the spikes. In some cases, this happens by increasing the rate at which the spike is rising; in others it happens by increasing the impact the spike has at whatever height it has reached at a given moment (remember that all four spikes continue to rise rapidly as you read this).

For example, water shortages increase the rate of extinctions by desiccating delta wetlands, or by cutting off tributaries to lakes. In central Asia, the Aral Sea between Kazakhstan and Uzbekistan has been so dried out in this way that great seagoing ships now lie on barren sand miles from the shoreline, where they were stranded like beached whales when the water receded. The shortage increases the impact of a given level of population spike by shrinking the food supply to that population. [The immediate effect may be to cause the spike to fall back as people starve, but that prospect only sharpens the signficance of the spike: that if we don't bring it down by carefully and humanely planned means, it will come down of its own accord, in ways no humane person would wish.] When the Aral Sea was robbed of its water by upstream withdrawals from its tributaries, its great fishing industry, which in the 1950s had been producing 100 million pounds of marketable fish per year, was completely wiped out. And the shortages increase the impact of a given level of consumption by reducing resource reserves. As reserves decline, we creep closer to the point where any more consumption by one community means more starvation for another—and where disaster can no longer be averted.[1]

To the extent that world leaders have awakened to the severity of water shortages, it's the direct conflicts resulting from inequities of allocation that have galvanized them. In addition to the 26 or more countries that are now water-deprived, hundreds of additional regions within countries, including large areas within China, India, Mexico, and the United States, among other high-impact nations, are water-scarce. Many treaties and cooperative arrangements have been set up for river basins that are shared by more than one country. These are critical, because there are growing signs that conflicts over water could be a major cause of future wars. And significantly, these water agreements cover some of the world's most volatile places—the Jordan River between Jordan, Israel, and the Palestinian West Bank; the Indus River near the border between India and Pakistan; and of course the Nile, which all told involves *ten* African nations.

Conceivably, some jurisdictions will alleviate their shortages by getting smarter about water efficiency, as Israel has done with its advanced drip-irrigation techniques that greatly reduce evaporative loss and waste. But the overwhelming temptation, as thirst quickens, will be for those who are nearest to the source to

1 ❧ Scarcity of water for irrigation has sharply curtailed our ability to make increases in food production keep up with increases in population. As a result, total global reserves of grain have fallen from a more than 100-day supply (or "carryover" from one year's harvest to the next) in the late 1980s to about 50 days in the late 1990s. That means the world now does not have enough food in stock to see it through a whole winter. The northern regions now must depend on importing from lower latitudes or the southern hemisphere during the late winter and early spring.

grab what they can—the agreements be damned. Aside from the escalating risk of war, such moves to protect upstream farms and ecosystems would only hasten the demise of the downstream ones, and overall biodiversity loss in the basin would only worsen.

More importantly, with policymakers focusing on the question of who gets how much water, they are almost entirely overlooking the prospect of pervasive secondary or synergistic impacts, such as sharp changes in the prices or availabilities of food, fiber, or wood, or in the climate itself, which will affect every water-dependent person, plant, or animal on Earth. Ethiopia might try to reduce the flow to Egypt, for example—but that will not prevent Ethiopia from suffering from devastating transboundary impacts.

One of the most disruptive of these impacts is that of overall water supply on the overall human food supply—in which Ethiopia has a huge interest because it will be so heavily dependent on being able to import grain from somewhere else. In this respect, the future will be radically different from the past, when one region could effectively insulate itself from another. Throughout the twentieth century, people in affluent countries were fully aware that people in poor countries were suffering from sporadic famine, but it was always assumed that the anguish was too far off to reach them. That is changing fast, though not quite in the ways we might have expected.

When a study by Lester Brown of the Worldwatch Institute calculated that China's demand for imported grain would exceed the entire world's supply by the third decade of the twenty-first century, the disconcerted reactions of UN, US, World Bank, and Chinese officials showed how blindered traditional methods of forecasting have been as a result of the specialization and fragmentation of knowledge. Agricultural economists had not been paying attention to the paving-over of land their extrapolations assumed would be available. It had apparently never occurred to these specialists that the amount of arable land in China might be changing.

It had also never occurred to them that the amount of fresh water with which to irrigate that land might be changing—and changing fast. Water is fundamental to all life, and all life competes for it. Whatever human institutions are to successfully shepherd our progress through the twenty-first century will have to capably manage that competition for water. China's government rules one of the oldest cultures on Earth, and it is a culture whose governments have long been aware that they could rise or fall depending on how well they managed their water. More than a thousand years ago, the Chinese began building a 1,200-mile-long Grand Canal to carry food from the rainy south to the arid north. The Chinese also built hundreds of reservoirs, dams, and canals, and drilled hundreds of thousands of wells. No people had ever been more water-conscious. Yet Brown's analysis suggested that as a nation, China had been blindered in a way that could have proved—and could yet prove—catastrophic.

The competition for water takes place on a number of levels. In direct consumption, it's a competition not only between nations but between economic "sectors" within nations—between farms, industries (which use water for cooling and processing), and cities. China has not had severe water disputes with other countries, like those occurring in the Middle East, since its agriculture, industry, and population are all concentrated on the eastern side of the country facing the sea, not in the mountainous areas fronting other countries. So, its recent concern has been with its own internal competition for water—between the city of Beijing and the farmers around Beijing, for example.

The competition for water is not only essential for direct consumption, however, but as part of an economic equation. If water is a critical input to industry, restricting the supply to factories could impair the country's ability to produce goods for trade. And, if water is a critical input to urbanization (per capita use in cities, with their modern showers and toilets, is much higher than in rural villages), restricting municipal use would cripple the urban growth that is essential to industrial growth—not to mention triggering politically threatening protests from the rising middle class. So, the Chinese government, intent on securing its place as a great economic power in the world, has been loathe to curtail the availability of water to its factories and housing developments.

What Brown's analysis showed was that as industrial and urban water extraction increased, the water available to agriculture was falling fast, and that this would impact on another kind of competition—more indirect, and still a few years over the horizon—that the Chinese government was not taking into account.

Brown had been intrigued, in the early 1990s, by reports that the Yellow River, the most important source of irrigation water in China (and one of the two most important in the world) had fallen so low that it became nothing more than a trickle by the time it reached Shandong Province, one of the country's most productive agricultural areas. Shandong sits astride the Yellow River Delta and depends on the river for the bulk of its irrigation. But a quarter century ago, the river had begun to falter. In 1972, it had dried so severely that it failed to reach the Yellow Sea for 15 days. By 1985, it was running dry every year, with the dry stretches getting longer. In 1997, it had gone dry for 226 days.

But this wasn't just an anomaly of the Yellow River. If you look at a map of North China, you'll find the city of Taiyuan about 400 miles south-west of Beijing, on the North China Plain. In the *Rand McNally New International Atlas* published in 1995, the Fen River can be seen flowing south through Taiyuan to the Yellow River. But by 1998, the Fen no longer existed. A large amount of water had been withdrawn to run Taiyuan's thriving coal industry, and drought had done the rest. Few people in that region, I would guess, had paused to consider how the smoke from their coal might be contributing to the global warming that exacerbated the drought that dried their river.

The city of Taiyuan, which with 2 million people is about the size of Paris, will either have to be abandoned or have its water transported at huge cost about 250 miles, uphill, from the Yellow River. The trouble is, as noted, that the Yellow River has nothing to give. Meanwhile, over 300 of China's largest cities, despite having first shot at the water (ahead of the farmers), are now experiencing scarcities. Not surprisingly, millions of farmers have had their wells go dry. A survey by Liv Yonggong of China Agricultural University in Beijing estimates that the water table under the North China Plain has fallen by about 25 feet in the past five years.

Brown's concern was not that people in China would starve, although starvation had been a recurring specter in that country for millennia. In the 1990s, China's economy was booming, and if there were impending food shortages China could probably afford to import surpluses from other countries. The problem lay in how much China would have to import. His analysis showed that by 2030, China would need an extra 200 million tons beyond what it could produce itself. That 200 million tons happens to equal the entire export capacity of the world. If China were to satisfy its burgeoning demand over the next quarter century by using its new money to corner the global grain market, that would leave no grain for any other country to buy. But meanwhile, the growing populations of grain-poor countries around the globe—such as Ethiopia—will have vastly increased *their* needs for food imports. The UN projects that the North African countries of Egypt, Sudan, and Ethiopia will have nearly tripled their populations between 1995 and 2030. Yet, those countries were already using 100 percent of their irrigation capacity and still having to import food. So, all the increase in population between 1995 and 2030 will add to the global demand for imports. Similar increases in demand will come from such population-spiking countries as Afghanistan, Burkina Faso, Burundi, Liberia, Mali, Mozambique, Nigeria, Pakistan, Saudi Arabia, Somalia, and Tanzania, nearly all of which are projected to have doubled within the next 25 years.

Meanwhile, however, the world's overall food production capacity has leveled off, and on a per capita basis has actually been falling, for nearly a decade. The ideological contrarians argue that "technological advances" will increase production capacity. Their refrain is taken up by the information launderers and percolated through global consciousness. But technology is not magic; it is always constrained by the laws of nature. Agricultural technology can increase yields by redistributing the "phostosynthate" (the material a plant can make from a given input of sunlight, soil, and water). Scientists can, for example, re-engineer the plant so that more of the photosynthate goes into kernels of corn and less into leaves, stems, and roots. But there are absolute limits to how far they can go with this, since a plant with weak roots would die in the heat, and one with a weak stem would fall down in a wind—yet both heat and winds are likely to get harsher. In short, when fully fertilized (as most of the world's farmland now is), a plant can only produce a certain amount of food from a given supply of water and soil. And both of those inputs are falling, as water scarcities are exacerbated by soil erosion.

The bottom line, said Brown, is that even if modest increases in yield are achieved, the large population growth yet to come will consume those increases and more. Even optimistically assuming the world's import demand can be held as low as the 200 million tons it was in 1998 (and it probably can't), satisfying China's demand could leave hundreds of millions of people without food.

For those affluent who are in the obsolescent twentieth-century habit of assuming this still only affects poor countries, Brown cryptically pointed out that such massive scarcities would likely trigger "social disruptions." Lester Brown doesn't like to draw a graphic picture of what such disruption might mean, other than occasionally mentioning the prospect of "food riots" and "destabilization." Perhaps even for someone as unblindered as he, there's a certain security in not venturing beyond the boundaries of quantifiable trends.

But some of his colleagues went a step further. Worldwatch senior researcher Michael Renner, analyzing what happens when countries are severely resource-deprived, wrote that such disruptions constitute a sea change in the nature of what was once called "national security." This concept of the security of the state became a central preoccupation in the planning and spending of governments over the centuries in which the main threats to a nation were the military forces of other nations. Renner and others pointed out that in the last two decades of the twentieth century, while the military planners were spending billions scanning the skies for missiles that no one could afford to launch (and that the main potential launcher of which no longer even existed as a political entity after 1989), the threats of food and water scarcities, overcrowding, and disease had crept up behind them.[2]

Renner did not learn until 1998 that four years earlier, the US National Intelligence Council—the almost invisible umbrella organization that coordinates US security issues—had read Brown's articles in *World Watch* and decided to launch its own analysis. The NIC was particularly concerned that the *World Watch* series (which I edited) diverged sharply from extrapolations of global food capacity that the US government had relied on for decades. The NIC assembled a research team of prominent scientists under the aegis of a CIA project called MEDEA, the aim of which was to confirm or refute the *World Watch* warning. The MEDEA team monitored China by satellite and collected data on cropland and water. In 1998, the report was completed, and its conclusion closely confirmed Brown's projection of China's import needs: China would need to scour the planet for an additional 200 million tons of grain per year in 2030—equal to the whole world's export capacity in 1997. That would indeed leave about 160 other countries short, some of them catastrophically short.

The MEDEA study was unnoticed by the mainstream news

2 ❧ At one point in the summer of 1998, the three largest armies in the world were engaged in activities no conventional strategic planning is likely to have anticipated: the US army was being inoculated for anthrax, a threat posed not by other nations' armies but by individuals against whom armies are not equipped to defend; the Chinese army was frantically wading through chest-deep water to repair dikes against the flooding Yangtze River; and the Russian army was being sent into the forests to gather mushrooms and berries for food, in lieu of pay.

media, but it may turn out to have been a turning point in public perceptions, in two respects. First, it could signal a sea change in understanding what security really means—that it's no longer mainly a protection against doomsday weapons or a communist conspiracy, but against something both as ancient as the species and shockingly new. Second, it signals that the potential for famine is no longer an isolated phenomenon in a few particularly misfortunate places, but a large-scale likelihood for scores of countries with hundreds of millions of people.

In all likelihood, there are people living in rich countries who are still inclined to think—even with hundreds of millions of their fellow humans now in desperate straits—that the problems will always be somewhere else. But that inclination was conditioned in an earlier era, when national economies were far less interdependent. Today, a disruption in one country has instant ripple effects in scores of others. Lester Brown told me a story he'd heard in 1998, about a farmer riding a large modern combine in Kansas. The combine was equipped with a radio tuned into international business news via satellite, and as the farmer operated the machine he heard a report that because of the severe water shortages, China's grain imports would rise that year. The farmer whooped with glee. In its effects on short-term US export income, this was good news for him. At this point, China's imports were not yet visibly deepening the deprivations in other countries.

But the Kansas farmer's reaction underscores a critical point about fresh water in the twenty-first century and forevermore. We can no longer deceive ourselves that water resources are strictly the assets of some regions but not others, because water can be transformed into grain, which is a tradeable commodity in every country on Earth. It takes 1,000 tons of water to produce one ton of grain, and while the weight of water may be too prohibitive to move by mechanical transport, the weight of grain is not. (Of course, grain can't be used to take baths in, or to drink. But the biggest user of water is agriculture.) Water, in other words, is not only physically but economically liquid.

A massive water shortage, then, will drive up prices of grain and everything grain is used to produce (beef, chicken, pork, farmed fish, eggs, milk, cheese, and beer, as well as cereal, pasta, and bread) in every country in the world. In poor countries, food could be afforded by the wealthy elites, but those elites could quickly become targets for millions of hungry poor. Now return to the scenario of an Egypt caught between the pincers of declining water supply and exploding population. For the sake of coherent discussion, that scenario isolated the potential effects of events in a few countries. But in fact, the real-world prospect is that all countries will be drawn into this maelstrom. Leaders, unable to deliver relief, would likely be toppled, and governments would lose their ability to maintain order. Millions of people would depart as refugees, spilling over borders in diasporas too large to either control or support. The influxes would further erode national borders and identities, and many of the flows would head for Europe and North America.

In the wealthier countries, as in China or Indonesia, the affluent could afford to pay the inflated food prices—but many of them too could find themselves becoming targets. In the United States, for half a century, the wealthy have been given a free pass by the poor, thanks to the myth of the "American Dream." That dream was given a final spurt of life by the era of entertainment-dominated consciousness, in which we saw case after case of stupendous upward mobility—poor kids with poor educations making $5 million-per-year salaries by age 21; Appalachian housewives becoming rich celebrity singers. That was supplemented by $100 million lotteries, "You have just won $1 million" sweepstakes, and other highly publicized means of convincing the poor or struggling not only that they could realize the Dream, but that they might even do so overnight.[3] Arguably, people didn't want to bring down a regime that might soon be theirs. They liked the consumption spike and wanted to ride it into the heavens.

But when food itself becomes hard to come by, such fantasies can be quickly overridden by hunger pains—and growing resentment. (When cooking oil became scarce in Indonesia in the summer of 1998, the resentment rapidly escalated into street riots.) The United States and Europe already have embarrassingly large numbers of poor, and by the late 1990s the distinction between "developed" and "developing" countries was already blurring. The US and European poor may live in drab apartments or alleys, or even under bridges, but as long as world food prices are cheap, they remain relatively quiescent and invisible.[4] With prices they can't afford, however, the poor could become loose cannons in the capitalist economy. Among the possible outcomes: demonstrations on the front lawns of the wealthy; Robin Hood-style raids on supermarkets, restaurants, and the kitchens

3 ❧ Eric Brown, of the Maryland-based nonprofit Center for a New American Dream, notes that TV programming consistently depicts "typical" people as much more affluent that the average viewer really is—and instills unrealistic aspirations and spending habits. "Americans are fighting and losing an expensive battle not with their neighbors across the street, but with the rich and the famous. Goodbye, Joneses, hello, Bill Gates," writes Brown.

4 ❧ The United Nations reported in 1998 that more than 100 million people in the rich nations are homeless, and more than 200 million are so deprived of food or medicine that they are not expected to live to age 60.

of the affluent; hijacking of food delivery trucks; escalating demands for increases in minimum wages, food subsidies, and other social safety nets; rising backlash by the haves against the annoying agitation of the have-nots; hoarding; and impossible pressures on politicians from opposite sides. The net results could be a dangerous polarization of societies in which economic disparities have been quietly widening for years, and destabilizations that many governments could not survive.

We have little time to solve this problem of water scarcities that begin in the Yellow River basin of China or Nile headlands of Ethiopia, but spread quickly to the once protected enclaves of the well-to-do. Moreover, there won't be time for big mistakes, because the road to a 200-million-ton grain deficit in 2030 could easily mean passing through a deficit in the tens of millions within the next few years—enough to trigger severe disruptions in civil organization. One of the biggest conceivable mistakes would be to rely heavily on the agendas of autonomous nations, both because those agendas will often be in conflict and because as viable institutions, many of those nations are on the endangered list. In others, such as the United States, the policy agendas have become so absurdly distorted by disinformation that it would be foolish to assume they will prevail much longer. This will not be the first time everyone on Earth has had a common interest in taking cooperative action, but it may be the first time that that interest has been visible to most.

To devise a water-management strategy that won't crumble into feudal disputes between outmoded nations will require the same broad, transnational approach that's needed to stabilize the spikes. That's because water scarcity is just one manifestation (though a particularly critical one) of the breakdown the spikes have brought. The same point can be made about inequities of wealth, resurgences of disease, the rise of killing, or the loss of human cultures and languages. In each area, the solutions are overwhelmingly of a nature that national governments can't solve alone and in some cases are only likely to make worse. In other words, do what we have to do to stabilize the spikes, and the policies needed to solve these other problems—including water management—become much clearer.

- The carbon gas spike can be slowed to some degree by trapping carbon gas in trees and plants, as is normal for all plant life. The problem is that as heavy mining of water for industry or municipal use leaves marginal lands with insufficient supplies, cropland is abandoned and often turns to desert—cutting down on the amount of carbon that can be sequestered in plants or trees, and thus raising the spike still faster. About 200,000 square kilometers of land turns to desert each year. Over a decade, that's more than the area of California, Japan, and England combined. An even larger amount of land is deforested in ways that may not result in desertification, but do continue to shrink the carbon sink. Much of that is in watersheds shared by two or more nations.[5] Moreover, the deforestation in these nations is continuing relentlessly. As the synergistic effects

5 ☙ The Ganges River Delta, for example, is shared by India and Bangladesh. About 85 percent of the Ganges has been stripped of its forest, and one result is that the natural flooding of the river during monsoon season has greatly worsened (when tree roots and forest detritus are removed, the water moves much faster over the ground). Most of the river flows through India, but Bangladesh takes the brunt of the flooding; in the summer of 1998, two-thirds of the country was inundated—including the megacity of Dhaka. Bangladeshi resentment toward India is festering.

of the spikes have shown, such deforestation is a problem with many faces: it reduces the capacity of the watershed to filter groundwater; it speeds erosion; and it exacerbates flooding. The policies that stabilize water management, therefore, by restoring tree cover will also trap carbon. But if a watershed is shared by several nations, they may be helpless to make this happen on their own. The logical unit of governance, then, is not a nation but a watershed district.

- Extinctions can be slowed by halting the devastation of coastal estuaries, which are among the planet's richest ecosystems—but also the most endangered. Rivers are the primary sources of water for cities and industries, and as those users draw more water, the lowered water downstream dries out those estuaries. In the Colorado River basin, of the 50 native fish species that thrived a century ago, 29 are now either endangered or already extinct. One reason is that the Colorado flows through both the United States and Mexico, and the two countries can't agree on how to manage it. National governments also tend to perform poorly in the protection of biodiversity even in the rivers and lakes within their boundaries, because their fixation on global trade often causes them to sacrifice ecological needs for those of industry, as is happening in the rivers of China or in the mangrove forests of Thailand. There may be a need to shift power from nations both to watershedwide institutions (for rivers crossing international borders or covering large regions) and to local communities (for smaller watersheds or ecosystems). Nations will need to let go of their central control and let more of the stewardship of local ecosystems return to local people—perhaps under the management of the bioregional institutions, coordinated by global institutions rather than by nations. Hindu nationalist prime ministers, Muslim militants, US western senators, and Thai shrimp mafia will be out of the loop.

- Population growth has historically centered around rivers and other well-watered areas, and societies' capability to support that growth without collapse has depended on their ability to produce enough food surplus to support the variety of nonfarming occupations that make up a diverse and thriving society. In other words, the extent to which a population spike of 8 or 10 billion is able to avoid catastrophic collapse depends in part on the efficiency with which it can convert water to food. The higher the yields, the higher the population that can be supported. The danger is that the higher the population climbs, the greater the crash will be if that efficiency can't be sustained. It's like watching an acrobat climb a vertical pole with his arms: the stronger he is, the higher he can climb—and the more risk he takes that when his strength finally gives out he will fall.

Thus, as higher population diverts more water to showers, toilets, and factories, the increasing scarcity to agriculture will drive up the efficiency with which we use water, through increasingly sophisticated techniques of irrigation and recirculation—but will also make us increasingly vulnerable should those highly tuned systems ever fail. Unfortunately, the need to use river water as efficiently as possible is blatantly undermined by the inclinations of individual nations to gain as large a share as possible. Why would Egypt rush to reduce its use of Nile water through greater efficiency, if it thinks that will give Ethiopia grounds for demanding Egypt's share be reduced?

- The consumption spike is the one that nations are probably least capable of stabilizing, because their conceptions of national identity, and even survival, are so tied to the idea of "development" (as in "developed" nations versus "developing" ones). Economic development, as presently practiced, is aimed explicitly at increasing consumption—both domestically (to raise household income and GDP) and abroad (to raise export income). Some of this practice is hard to fault, at least for those of us who find it hard to imagine life without showers and flush toilets. In the developing nations, as the number of people who can afford bathrooms rises, so does per capita water consumption—dramatically. More efficient designs can mitigate that somewhat. But those designs do nothing to cut back on the huge amounts of water consumed by the shift toward more meat consumption, and toward higher consumption of products like paper and leather and plastic (all of which require huge amounts of water to produce), as development proceeds. The more urbanized, industrialized, and modernized we become, the more water we consume.

Because a rising share of the consumption is in products that are traded internationally, this means—in effect—that water shortages in one part of the world will be felt in many other parts not only through the distribution of food, but through the sales of hundreds of other products. And, while the control of water-as-food may have the more primal impact, water-as-consumer-products may wield the greater financial clout. The reason is that in the current global economy, the market value of water to industry is far greater than it is to agriculture—which is one of the reasons so much water has been diverted from farms in the first place. A thousand tons of water used to produce a ton of wheat, for example, has a market value of $200, while the same amount of water used in a heavy industry yields, on average, about $14,000 in output. As long as nations are in competition with each other to capture export income, that industrial stranglehold on water and the raising of the consumption spike—will continue.

On a technical level, it's not hard to see how these exacerbations of the spikes by water scarcities—and vice-versa—can be stemmed. The carbon sink (which could someday begin to pull the emissions spike down if fossil fuels are phased out) can be expanded by increasing forest cover, but that can't be done without watershed protections that have far stronger teeth than the saws of the farmers, developers, and industrial timber industry have. But that means protecting whole watersheds, not the pieces of a watershed that lie in particular states or countries where the exploitation by one might undo any conservation gains by another. Extinctions can be stemmed by stopping habitat destruction and bioinvasions, and that too means setting policies for areas defined by natural borders rather than political ones—not just the ridge boundaries of watersheds, but the coasts of the oceans and seas, the timberlines of mountains, and the moving boundaries of vegetation zones.[6]

6 ❧ The time may come when our maps of the world no longer show fixed boundaries of nations printed on paper, but continuously fluid boundaries of bioregions tracked by a global network of monitoring stations and satellites. The jurisdictions governing human actions, too, could be fluid and responsive both to the day-to-day challenge of climatic disruption (dealing with hurricanes and floods will have to be routine, not "emergency"), and to the longer-term needs to accommodate migrations, bioinvasions, and changes in water tables or forest cover.

Whether in the management of fresh water, oceans, wetlands, or forests, achieving stability will require both more efficient use of resources and reduced absolute consumption. Reduced consumption will require shifts both in human perceptions of what satisfies, which now drive the demand for dubious acquisitions, and changes in the efficiency with which we achieve that satisfaction. These changes will very likely mean more power for real, rooted-in-the-earth human cultures and less either for governments that create jealousies over artificial boundaries or for transnational corporations that create artificial appetites.

That's not to say that salvation lies in renouncing national governments altogether—at least, not now. The famous assessment of democracy ("it's the worst system on earth except all the others") might apply just as well to nations, to the extent that they act as geographically large units of human organization. They still serve important functions, not the least of which is to collect taxes and provide funding for the protection of the commons. Governments can provide some of the monitoring and mechanisms essential to assuring human security, and the assurance of access to information that protects people from tyranny—whether it be the tyranny of industry, media, shadow economy, or government itself.

What we do need to encourage, however, is a revolutionary shift in the way government goes about protecting security—a shift from maintaining massive military defenses and demanding arbitrary allegiances to mounting serious biological, climate, and population-growth defenses.

8

YOU

❧

When the Chernobyl nuclear accident happened in Ukraine, it most likely didn't affect you directly. The same might be said of the Bhopal toxic accident in India, the Exxon Valdez oil spill in Alaska, and so on down the list. If you were directly hit by any of these, you are one in a million.

In this respect, the global spikes are different; they will not pass you by. They'll touch you in ways that are predictable and in ways that will take you quite by surprise. They will change your life profoundly. They may change where you live, what work you do, how you eat, and how you think.

Not everyone will be affected in the same ways. You may or may not live in a flood-vulnerable basin, a fire-prone area, a drought zone, or a water-scarce region. Or you may live in such a place and not realize it. But chances are high that you live in one or another, or that you will before long, because perhaps half the world's people live in such directly spike-vulnerable areas, and those areas are continually expanding. In China, when the Yangtze River basin flooded catastrophically during the summer of 1998, there were more people living in the Yangtze watershed than live in the entire United States. About 56 million of those people were flooded out of their homes. During the same summer, 21 million were flooded out in Bangladesh.

At least 54 countries experienced major flooding in the late 1990s, and at least 45 were stricken by severe drought. Of those, 22 countries experienced both floods and drought. And these were not just poor countries in the far reaches of the African Sahel or the Indonesian archipelago. In the United States, fires swept large areas of Florida; drought desiccated Georgia; and 29 straight days of 100-degree-plus weather staggered Texas—driving the Ogallala closer to extinction. More of Earth's surface was scorched by wildfire in those years than ever before in recorded history.

If you live in any of those places, the risks of disruption are rising. And *wherever* you live, you'll be more vulnerable to infestations and resurgent diseases, angry weather, and economic ripple effects of events elsewhere on the planet. You may have had inklings of this, but may also have concluded that there's little you can do—that such changes are far too overwhelming and enveloping to be worth fighting. That conclusion would be mistaken.

Chances are, if you live in an industrialized country you have planned and managed your life so far to a considerable degree: you've invested in education or training; you've put money into business enterprises or financial investments; you've bought insurance against various kinds of losses. Imagine how much more helpless you'd be if you had not gone to school, saved money, or had insurance when your car crashed or you needed medical care. Now the same kind of logic applies to the threats of the spikes, only more so. In a world where your life could be thrown into an abyss by a catastrophic illness, a house fire, or a car crash, life has always been a game of Russian Roulette; while many of us are wounded, many more are not. In the world of the spikes, however, all chambers are loaded. The time has come to plan your responses with at least as much focus as you've ever given to your education or finances.

It's helpful to think through your responses on three levels: your personal life, your community life, and your role in the larger economy. There may also be a fourth level, which might be described as the role you choose to play in the next great transition of civilization. That transition is the path we will take collectively, from the unstable hot-house growth that characterized the past century, to whatever more stable form will follow. On each level, we can identify a strategic goal that gives you the best chance of surviving and even thriving, while also best serving the common needs of humanity. It should be clear by now that the two are inseparable.

PERSONAL SECURITY

As far as we know, the main threats to health or safety will come mainly from climatic and biological disruption: from traumatic storms, floods, contaminated water or food, disease-causing insects or rodents, heat, and fire. In most cases, the most effective defenses will come through strong community efforts and political activism. But there are steps you and your family or friends can take as well.

Bad water: Contaminated water is already the world's largest disease carrier. The synergistic effects of climate change could worsen that—by damaging infrastructure (allowing sewer water to back up into drinking water supplies), by increasing water temperature and extending the ranges of pathogens, and by drying up ground water sources and forcing more people to draw water from rivers or streams that have not been filtered the way spring or aquifer water has, and are more likely to be polluted.

Wherever you live, simple precautions are now warranted.

Know the source of the water you drink or bathe in (pathogens like pfisteria can be carried by skin contact as well as drinking) and whether it has been treated, and use water that has been filtered, distilled, or bottled at a spring if you can. If you have any reason to suspect biotic contamination, boil the water you're going to drink. In the years to come, always keep an extra week's supply of bottled drinking water on hand and keep rotating it.

Extreme storms: Areas like the East Coast of the United States and the Caribbean that have regular hurricane activity could experience a sharp escalation. Scientists estimate that sea levels will rise as much as 3 feet over the next century, and with every inch an estimated 2 million more people become vulnerable to storm surges. In low-lying regions like Bangladesh, vulnerability to monsoon-driven floods and storm surges will increase. In river basins downstream from major mountain ranges, like the Ganges River in India and the Indus River in Pakistan, snowmelts may be larger and could cause inundations. Low-lying coastal or delta cities, such as Dhaka, Calcutta, Miami, and New Orleans will be in growing jeopardy.

If you live in hurricane country, make sure the structure you live in can withstand strong winds (many inexpensive new homes can't). If it can't, reinforce the building or move. Check your insurance to make sure you have adequate coverage for "acts of God," even though we know such damage may now be partly human-caused. (The insurance companies are well aware of this and are taking care of themselves by adjusting rates in anticipation of increased damages. And sooner or later, the US government is likely to decide it can't afford to continue providing protection for people who live in the paths of storm surges or floods.)

If you own oceanfront or floodplain property, act while you can—don't just pass the property off on someone else, but try to mobilize your community and country to enact policies that can let you move without losing everything. The insurance companies, which know the changing odds, are withdrawing from coverage of such properties. This is where personal action has to bond closely with political activism: worldwide, there are hundreds of millions of people who live in flood-prone areas (400 million in the Yangtze basin alone, for example), and few can afford to abandon property. Yet, if they are forced out, they will become floods themselves—floods of refugees pouring into cities on higher ground. Whether you're a potential flood victim or a potential refuge-provider (almost everyone is one or the other), it's cheaper to begin spreading the cost of relocation now than to wait for catastrophe to take ever-greater tolls.

In the meantime, if you're in a vulnerable location, review your emergency procedures in the same way Californians are routinely advised to know what they'd do in an earthquake. Decide where you'll go for physical protection if a hurricane, flood, or tornado approaches with short warning; know in advance what important records or valuables you'd want to save and where they are; and if you retreat to a temporary shelter be sure to take your supply of drinking water. It would also be smart to have an emergency kit available: a backpack containing a portable light, phone, radio, knife, canteen, water purification tablets, carbohydrate bars such as those used by endurance athletes, and perhaps a waterproof bag for photos, valuable papers, or whatever you can readily carry that is important to you. In fact, deciding how to equip this kit could be a useful exercise in assessing what's really important to you: if the contents of your home would fill a truck, but you suddenly had to abandon all property except what you could carry on your back, what would you take? Don't wait until there's no time to reflect to decide.

Wildfires: In recent years, fires have increased enormously, and all four spikes are at work. Global warming dries out forests and grasslands (not so much from slight increases in average temperature, as from disruptions such as droughts and extreme heat waves); population and consumption pressures are driving the kind of slash-and-burn deforestation that has driven extinctions in the Amazon and Indonesia; and those fires are the kind that have raged out of control on every continent. Californians, who have been hit by forest or brush fires repeatedly, have learned practical precautions: don't build houses in chaparral forests, which burn like gasoline; don't live under a roof of cedar "shake" shingles, which burn like paper; don't build in a place where prevailing high winds like the Santa Ana would place you directly downwind from an adjacent forest. Californians are also advised (or required by law) to keep brush clear around their homes, but ecologically that's a rob-Peter-to-pay-Paul measure, because it adds to the fragmentation of the forest ecosystem, which can backfire in other ways. Better to pull back from participating in the development of such areas altogether. If you're thinking about where to live in a region like that, it's both ecologically and personally safer to settle in an already urbanized area.

Water or food scarcity: If you could afford to buy this book, you probably won't experience food scarcities imminently, though millions of others will. But you could soon see changes in what you eat and how much it costs. Meat production is heavily subsidized by its uncounted ecological costs and by its hugely inefficient use of grain, and as governments wake up to these costs the subsidies may be cut and prices of meat may soar. Prices of fish are subsidized by governments looking for export income, encouraging the overfishing that drives depletion and extinction. So fish, too, will become more expensive. Yet, within this problem of costly protein lies a great opportunity. Meat is as troublesome for human health as for ecological health, and cutting back on meat consumption could turn out to be far more of a benefit to you than a hardship. Numerous studies show that diets low in meat but high in grains, fruits, and vegetables (all of which require far less land and water than meat to produce) are more conducive to health and longevity. In an era when diseases are on the rise worldwide, and dependence on antibiotics is becoming less reliable, the importance of having strong immune systems and preventive practice grows higher.

A more indirect ripple effect of food scarcities in distant places, beyond raising prices, may be the rising reliance on bioengineered foods. Some scientists have raised urgent alarms about the risks that such engineering will have unintended consequences on the scale that pesticides or nuclear power did. One of the risks, for example, is that crops engineered to be resistant to particular herbicides will leak the genes that provide that resistance into the weeds those herbicides are supposed to destroy—thereby giving rise to generations of more virulent weeds than ever. Another risk is that crops engineered to produce their own insecticides, such as Monsanto's Bt cotton, will spread their genes into the plants of the surrounding fields or woods, causing one of the true nightmares of technology gone haywire—toxic chemicals that reproduce.

Meanwhile, to obtain seed from the companies that now dominate production, farmers often have to sign licensing agreements that treat the seed as the seed company's property. The farmer is forbidden to use the seed for further breeding, or to save the best seed for future planting—the traditional practice known as seed saving. Agricultural researcher Brian Halweil notes that in France, farmers have expressed deep dismay over the dictatorial control large agribusiness companies like Monsanto now exert over farmers. "Genetically modified seed varieties often carry traits that necessitate the use of one or more agrochemicals," he writes. "In this way, an agribusiness company—often producer of both seed and agrochemicals—can integrate the sale of several of its products and command substantial control over the farming process. In addition to the seed, the farmer must purchase the fertilizer, pesticide, herbicide, and other inputs, without which the seed does not function optimally."[1]

On the other hand, if you are an independent-minded farmer and you choose not to buy the company's seed and pesticide, you could be wiped out by the newly ineradicable pests spreading from the fields of your neighbors who do buy the seed in order to get its short-lived protection. So, let's say you give up and go with the commercial product. At best you're caught in a chemical dependence that can only get more desperate as the pests or weeds again gain the upper hand (recall the rising incidence of suicides among farmers in China and India). At worst, the fear is, some genes from the bioengineered food—genes from a moth, or an eggplant, say—might work some mischief in you.[2] In short, we are entering a time of double-barreled threats to our food supply: first that there may be pervasive shortages, and

1 ❧ In frustration, in 1998, a band of 120 French farmers broke into a storage facility of the multinational company Novartis and destroyed 30 tons of transgenic corn seed. The protesters said the company had made them serfs on their own land.

2 ❧ Among the many engineered food organisms that have been approved for field testing or commercialization in the United States are potatoes with a waxmoth gene, cucumbers with a tobacco gene, tomato with a flounder gene, pepper with a virus gene, broccoli with a bacterial gene, apples with a silk moth gene, and potatoes with a chicken gene.

second that in the attempt to increase production as fast as possible, food industries will subject consumers to unacceptable risks of chemical or biological contamination. In the face of these risks, it now makes sense to obtain food as much as possible from local farms using organic (pesticide-free) and nonbioengineered practices. It may also be prudent to supplement the commercial food supply with a home or community garden. In cities, backyard or rooftop gardening can provide a valuable supplementary source of fresh vegetables, fruits, and herbs—along with the peace of mind that comes with producing your own harvest of food you know to be free of toxic chemicals, irradiation, or genes whose long-term effects on your body or environment remain unknown.

Information: As should be clear by now, contaminated information can be as dangerous as contaminated water or food. It can cloud judgment, feed addiction, dull the will to act, and induce depression or denial. Though last on this list, reliable information is perhaps the most important to your personal protection. You need reliable information to be *able* to monitor water, weather, and other aspects of personal health and safety. While it's necessary now to make good use of the Internet, TV, print publications, and other mediated sources, it's also important to remember that the communications technology are never anything more than extensions of our own human powers of observation. If the original powers weaken from disuse, the extensions become devalued and even dangerous—like putting the steering wheel of a speeding car in the hands of someone whose arms have atrophied. By all means read, listen to public radio, or explore the web, but at the same time beware of getting too disconnected from your physical environment, family, and trusted friends or contacts.

COMMUNITY

When we moved from hunter-gatherers to agriculturalists and city dwellers, we passed through a transition from which there can be no turning back. Because agriculture supports levels of population vastly beyond those of prehistoric subsistence, and we have taken the chance to live at these levels, we have cast our lots with community. In countless ways, we have chosen to be dependent upon the services of the community in exchange for the protections and satisfactions it provides. If we are now confronted by unaccustomed new threats, it is at the community level that we can act most effectively.

That's not to exclude the roles of national governments, which will be essential to enforcing such strategies as worldwide carbon emissions reductions, and regeneration of forests and oceanic fisheries. But national governments are only institutions; contrary to their propaganda, they are rarely the same as peoples, tribes, or ethnic populations, which have cultural identity and are defined by their communities. It's communities that generate the passion to defend ancestral lands, or sacred rivers or mountains. If you belong to a community that has traditions rooted in the land and reaching back in time, you have better footing for stepping forward into the future.

A community, by nature, is a group of people who help each other—provide mutual protection, trade skills, share common assets, and share some common satisfactions, memories, and aspirations. It has its origins in those ancient villages where the first farmers began growing wheat and building permanent settlements, so we have at least 7,000 years of dependence on community programmed into our brains. Arguably, the ability to participate in community life became a selective survival trait during that time. Loners were less likely to survive and pass on their genes to progeny who were part of a family or clan with a strong sense of their past and future.

In the distant past, the lone hunter could not share body warmth on a freezing night, or get secure sleep by taking turns with trusted companions watching for predators, or be sustained in the hunt by the knowledge that when he returned exhausted, others would be waiting to restore him with hot food, dry clothing, attention to his wounds, physical touching, and affection. Equally important, he would have no sharing of storytelling about past exploits or future hopes; there would be no legends or myths, or collective memories or aspirations. Without community there is only the present, which is fleeting.

If loners of the distant past were cut off from the protective structure of society, they may be even more so now. Whereas the rugged individualist of centuries past would presumably have had at least some rudimentary survival skills, today's would-be individualist likely grew up heavily dependent on industrial processing of everything. The neolithic wanderer or New World fur trapper at least had intimate knowledge of the trees, herbs, animals, and weather; the medieval peasant could coax calories and fibers from the soil. The modern individualist rarely has those skills; he is dependent on the very society he has rejected—a fact that may render him all the more resentful. The "survivalists" who retreat to their Montana cabins or Afghanistan mountain redoubts often depend on subsidies from others, or on theft or fraud, to obtain food, clothes, and vehicles that have been produced by more socialized people than they. In today's world, the word "survivalist" is an oxymoron: it refers to the kind of person least prepared for what is coming. The real survivors will be those who form the strongest communities.

As we move to the near future, our dependence on community will be greater, because while there are some actions you can take as an individual to deal with effects of the spikes on a defensive or emergency basis as noted above, the only way to actually restore stability is through communal and global action. There may be a kind of frontier justice in this: it was civilization that gave rise to the spikes, and civilization now has to pull them back down if it is to continue. The qualities that gave civilization its power to advance wildly out of control—specialization of knowledge, explosive accumulation of information, vast magnification of our original powers—now have to be harnessed to getting it back onto a sustainable track. It's a risky but unavoidable recourse. Like it or not, you are part of a global societal process now, as dependent on it as a blood cell is dependent on the process of oxygenation. The deal is, we do this together or not at all.

Unfortunately, a community is not the same as a city, a suburb, or a squatter settlement; any one of those could be (or contain) strong communities, but most do not. While the world has become more urbanized, it has become in many ways less socially cohesive. If your interest is in becoming part of a community that really has the capability to help bring life back to a sustainable course, you'll need to find a way of living and working among people who share some principles that are going to be essential to organized life in the twenty-first century. You can do this by finding a suitable community or by helping to revitalize a dead or dying one, or even creating a whole new one. However you go about it, the community will need the following attributes:

A sense of place: It's possible for community to exist independent of geography (people can be brought together under the umbrella of an Internet site, a magazine, a TV church), but the strongest communities have roots in real, physically distinctive places. The citizens are bound to a remarkable extent by their interest in the physical attributes of those places. Historically, that bond has been economically based: a community was formed around a fertile farming valley, where the celebrations of life focused on the rhythms of the growing seasons and the harvest, or around a fishing village, or a grassland where sheep could graze and distinctive wool products could be made.

Contemporary sensibilities about place may as often seem to be psychological or aesthetic, as economic. But those sensibilities may still have ancient ecological roots. In any case, restoring ecological connection is now critical, because it's in part the separation of daily life from what sustains us that has made us so easily blinded to the spikes. If you live among people who are conscious that their freshwater is produced by a cycle that includes filtering by forested watersheds, for example, they'll be less likely to approve suburban expansion that replaces watershed forests with housing developments. If they're conscious that a living ecosystem can react to inflicted injury the way an animal with teeth and claws might, they'll be more aware that a denuded river basin is far more likely to flood.[3] An environmentally rooted community will take active measures to prevent large insults to its natural environment. People who have a love of place may have more reason to work together to protect its integrity—and thereby nurture their own.

A sense of trust: In troubled times, some people will barricade

3 ❧ Wallace S. Broecker of the Columbia University Earth Observatory in New York, in a warning that climate change could come suddenly rather than gradually, commented in 1998: "The climate system is an angry beast and we are poking it with sticks. We don't know whether it's going to pay attention to the pokes. But if it does, it might rise up and do something we don't like."

themselves; others will go into denial. The changes we need to make, to succeed in our quest, are so great that many people simply will not make them—they'll be like the flood victims who refuse to be driven from their homes and end up drowned.

But if you decide to act now, you'll not be alone—and that will make it easier. There are already millions on the move—either seeking more rooted and enlightened communities or forming them. The place where you live is thus important both ecologically and socially; we need physical connection with the land that sustains us (thus the great need of city dwellers for parks and trees), and physical proximity or contact with people who sustain us. We are still mammals, and mammals have to touch. A group of people who care about each other will have a far better prospect of thriving in any environment—ravaged or pristine—than an isolated person or family. In the event of materials or energy scarcities or property destruction, you can share or rent with more confidence; you are not unknown risks to each other.

Ability to be a part of nature, not an imposition on it. As you seek a community or try to build one, keep in mind that the community can't work if it is built as a fortress against nature's rages, or if it is out of balance with the ecology that supports it. At present, most urban settlements are out of balance; they suck huge amounts of resources from the surrounding area, and expel huge amounts of waste. A healthy community is one that replicates the functions of a healthy organism on a large scale. If technology magnifies what a particular operation of a person's hand, eye, or cerebrum can do, community does vastly more—it magnifies what people can do, with all their faculties and in all their diversity of learned as well as inherited capabilities. But people can't do anything at all without the continuous support of natural processes; nor, for long, can communities. Healthy communities, and the economies that run them, have to be a part of nature.

YOU AND THE ECONOMY OF THE FUTURE

You've thought through your situation regarding the direct threats of extreme weather, fire, flood, and so on, and while you can't predict what will happen, you're aware of how to achieve the best security you can for yourself and your immediate family or circle. Let's say you've also recognized the power of a larger community—a population of people who share a sense of place and a willingness to trust in each other. You have your plans for personal security and community support. But is this enough? Is protection from disaster all we can hope for? There may be an inclination, in the experience of waking up to what's happening and preparing for it, to see the future through gray-tinted glasses. We may subconsciously recall the post-apocalyptic movies we've seen—*Water World*, or *Blade Runner*—in which the world is cast into a kind of nihilistic hangover after the binge of hubris (and, now, overpopulation and overconsumption). We may wonder if our future will be like Rome after the Vandals sacked and burned it, or Easter Island after the last trees were gone.

That inclination may be conditioned by our still-reflexive tendency to connect our contentment to our consumption. In waking to the realities of a fragile planet, we may be taking the view of tourists who've just had a fling at a luxurious resort and now regretfully have to pack to return to drab reality, looking wistfully back over their shoulders. All our lives, advertising and PR have conditioned us to equate our highest aspirations with high consumption. Yes, we liked "roughing it" when we went camping as kids, but couldn't wait to stop at the Burger King on the drive home. On a larger scale, we note that people in the developing world who've been roughing it for centuries move as fast as they can to higher levels of consumption when their incomes rise—from rice-based diets to more meat, from bikes to cars, from cars to sport utility vehicles. We assume that it's human nature to want to have the things a higher income will bring.

But there's a fallacy in this assumption, and it's quite aside from the fact that if the whole world acquired cars and ate burgers at American rates, the planet would quickly self-destruct. Even if 3 billion internal combustion cars and greatly increased quantities of meat could be produced with no ecological cost at all (they can't), there's a mistake in assuming this meets human wants. Alan Thein-Durning explains in his book *How Much Is Enough?* that when you buy a product, it's not the product itself that you're paying for, but the service or satisfaction that product brings. You don't buy a refrigerator because you covet a big box that's cold inside, but because you want to keep food fresh. You don't buy a car because you yearn to own 2,000 pounds of steel, aluminum, plastic, and rubber, but because you want easy personal transport. You don't want a burger because you'd like to devour the flesh of a subjugated species, but because your body hungers for protein.

Go down the list of things you'd like to have (or do have), and ask why, and it becomes clear that in most cases it's not the things themselves but the services they provide that are what you really want. When you pay money for stuff, it's really because that stuff brings you such services as mobility, cooling, heating, physical security, shelter, communication, sensory pleasure, satisfaction of hunger, alleviation of pain . . . and so on.

Now consider that for any service you can name, there are different ways that service can be provided—some of which take immensely more physical material or energy than others, or impose immensely more environmental or social damage. We learned a few years ago, for example, that our refrigerators don't have to be kept cold with chemicals that destroy the ozone shield. Consider also that the ways twentieth-century industrial societies have gotten their services are among the most wasteful or inefficient in history. The energy consumption of the average refrigerator in China, for example, was found in the mid-1990s to be three times that of the average refrigerator in Denmark. In other words, there's no one-to-one relationship between what we get out of life and what it costs. On the contrary, it's possible to live with the same services or even better versions of them, for a fraction of the ecological cost.

To do this is not just a matter of substituting a more efficient technology to produce the same product as before, though that will be part of it. There are at least four different kinds of strategies by which these changes can be wrought:

Product substitution: Replace obsolescent products with ones that use less energy or material. Replace conventional tungsten light bulbs with compact fluorescents, and you get electric light at a fraction of the energy cost—and it's cheaper in money as well. (The new bulb costs more but lasts much longer and uses far less power.) Replace your big car with a smaller one. Both of those substitutions reduce carbon emissions.

Materials recycling: Participate in community recycling programs and buy products made of recycled materials. If you work for a publishing, printing, packaging, or shipping company or any other large-scale user of paper or cardboard, consider that each ton of recycled fiber your organization buys helps to build the market for recycled fiber, which in turn helps to slow the deforestation that is a primary driver of extinctions.

Reuse: When applicable, this is even more effective than

recycling, because it eliminates the energy cost of remanufacturing (recycling paper saves trees, but still uses energy to pulp the old paper and remove the ink.) Reuse may require only an intermediary sales service (such as a second-hand shop) which involves almost no industrial energy use. Old bricks, for example, not only represent an opportunity to add character to new construction, but eliminate a hugely energy-intensive manufacturing process.[4] And reuse can often be done informally, as when your child receives clothes from an older brother or friend. Because children grow faster than clothes wear out, this is an ideal venue for reuse.

Borrowing and sharing: It takes only a slight attitude adjustment to make this a large efficiency gain, both personally and publicly. Many services are obtained from products that sit idle most of the time. Boats, lawn mowers, power saws, ice-cream makers, video cameras, and rubber rafts are among the thousands of things people buy that they use perhaps 1 percent of the time—if that. A typical American or Japanese household contains hundreds of such occasionally used products. This amounts to an enormous duplication of payments by consumers, and over-extraction of raw materials from the earth by manufacturers. Often, it's nothing more than a misplaced sense of pride (or perhaps the fact that you don't know your neighbors) that inhibits sharing of numerous products that would, with just minimal planning and coordinating, allow a group of neighbors to use far less "stuff" than they do now.

4 ❧ Recall that the ancient Indus civilization may have hastened its own demise by cutting its forests to fire the bricks with which to build its greatest city—thus exposing the city to greater risk of catastrophic flooding. In recent times, that stripping of forest protection has accelerated, as more than a billion people continue to burn wood for cottage industries or cooking.

Urban design: This is perhaps the least understood strategy, yet it has perhaps the greatest potential. Allowing cars and trucks to dominate modern transportation has caused cities to spread over farmland and forest with impunity, first in the United States but now in places like Thailand and China, where this emulation of western success is turning out to be a colossal mistake. Policy-makers with a modicum of vision have tried to address the pollution aspects of the problem by imposing emissions controls, and they've tried to alleviate the congestion aspects by adding car-pool lanes—or by building still more roads. But these are Band-Aids, just as the climate treaty with its 5-percent solution is a Band-Aid. Even the electric car is a Band-Aid. What all such attempts to "improve" traffic flow or emissions overlook—or avoid admitting—is that the real problem is the huge distances that have opened up in the past half-century between people's homes and their jobs, schools, shopping, and entertainment. Bring those things together into a compact, pollution-free and noise-free community, and the car is eased out of the picture. Envision a neighborhood where most of the places you have to go in the course of a week are within a mile or two. Most can be made by bicycling or walking, and for the few that can't, public buses are available. For the occasional times you need to haul larger loads or take longer trips outside your community, you may still own (or borrow, or share!) a car, but use it for perhaps only one-tenth or one-twentieth the mileage the average American does now.

This kind of urban design or redesign brings multiple new benefits. Since most of the people in such a community don't need to drive much, the community can give more of its land to parks and trees instead of pavement. The air is cleaner for breathing and more attractive to birds; the soil is cleaner for urban gardening and landscaping; the outdoor space quieter and less stressful. The community is more conducive to walking or bicycling, safer for children playing, and more attractive for such communal activities as outdoor music, soccer games, or conversation. And all such benefits are above and beyond the great reduction of CO_2 emissions, heavy materials consumption, and paving-over of suburban land.

This is not to suggest that you can single-handedly redesign your community. And even with political consensus (which is not now on the horizon in most countries), it would take years to retract existing sprawl—to pull back the boundaries and revitalize the core—of existing cities. But that's not reason to delay a fundamental shift in the direction new planning takes from now on. Every day, hundreds of square miles of forest or farmland fall to new development for which there would be ample space within the core. And the way that goes can be changed quickly.

In short, it's a completely unnecessary violation of nature to cut forest for new houses when obsolescent buildings or empty lots inside existing urban boundaries lie abandoned or depreciated and ready for rebuilding—when the replacement of car-clogged streets and highways with more efficient urban designs would open up enough space within existing boundaries to accommodate all projected population with less crowding, pollution, and congestion than now prevails. You don't have to make sacrifices of convenience, comfort, or safety to play a role in more beneficent community-building. Quite the opposite: under the conditions we now face, you'd be making those sacrifices if you chose to perpetuate—or try to perpetuate—the high-consumption lifestyle that created the bubble of artificial affluence we've been living in.

Nor is this strategy only an untested ideal. It's had at least a

30-year trial run. In the cities of Portland, Oregon and Curitiba, Brazil, for example, policies requiring compact development have measurably reduced the problems of sprawl and inner-city decay that plague most of the industrial world. In Portland, urban growth is contained within a boundary established by state law. Public transit is designed to reduce car dependence, thereby opening more space for parks, open plazas, and other spaces where people can relate to each other. The city's trains are designed for easy bicycle access, and 85 percent of new development must be with a five-minute walk of a transit stop. Mixed-use zoning allows people to live closer to their work, shopping, schools, or cultural life.[5] The result, in livability, can be seen in a comparison of Portland with Cleveland, a city that has followed the more typical development pattern of sprawling out over the surrounding countryside. Between 1970 and 1990, Cleveland's use of land expanded by 33 percent, even while its population declined by 11 percent, as residents fled to its perimeters. During the same period, Portland remained fixed in size yet proved attractive enough to grow in population by 50 percent.

Curitiba, which is similar in size to Portland, established a similar objective of minimizing car dependence—in this case designing its land-use around a system of radiating and concentric bus lines that make it easy to move about by a combination of walking, bicycles, and buses. As a result, during the past 25 years, the number of miles Curitibans travel by car has

5 ☙ Mixed uses also eliminate the deadening effects of large "single-use" tracts such as railroad yards, massive blocks of public housing, or highways that the urbanologist Jane Jacobs calls "border vacuums" that cut off the intermingling of neighbors and destroy the variety of experience much as monoculture shuts out biological variety on a farm.

declined by a third, even though the number of people in the city has doubled. And Curitiba has become regarded as one of the most livable communities in the western hemisphere.

At this point, you may be anxious about two questions. First, why you should bother to seek new strategies, when present commitments—or sheer inertia—may make it easier just to continue the style of life you're presently caught up in. Second, if you don't live in Portland or Curitiba, how you can realistically hope to find a place to live where you're close to your work and other needs.

The answer to the first question is that just continuing as you have in the past simply won't be possible much longer. Change is coming fast, and either you'll move toward it with some control or you'll be thrown into it. As for the second question, what Portland and Curitiba have done, any city can do. In fact, hundreds are already taking steps in that direction. And here's where your personal, community, and economic interests converge. On all levels, the same principles of sustainability apply: follow them, and you maximize your chances not only of surviving the coming ordeals but of experiencing a kind of triumph that few, if any, previous generations have known.

- Get to know the *place* where you live *as a bioregion*, rather than as a political jurisdiction. Prepare to defend it, not by giving your political support to the nation's "military preparedness," which is a largely obsolescent strategy for human security, but by vigorously opposing any development that would clear forest cover from the watersheds that serve that region, and by promoting policies that would curb road-building and sprawl. Get to know some of the key native species in your region, and the roles they play in the region's health—and in yours.

- Identify the end products of the *work* you do and assess how compatible that work is with a sustainable economy.[6] If it's a bad fit, think about a job change. If you think you can't risk a career change, think of whether you can encourage reform from within. For example, an oil company need not be a dead-end employer just because it's heavily invested in fossil fuels. It can reposition itself as an energy company and begin shifting smoothly into solar, wind, or hydrogen energy.[7]
- Live where you can use as *little fuel-burning transportation* as possible. If your past inclination has been to live in the suburbs because you like the greenery, consider that that greenery is coming under increasing pressure from fragmentation and bioinvasion, and many people are coming to realize that instead of fighting daily battles with the bug spray and commuter traffic, it's easier—and more rewarding—to live closer in and get their contact with nature on occasional weekend trips to rural retreats.

6 ❧ In 1987, Humboldt State University of California initiated a "Graduation Pledge Alliance," encouraging graduating seniors to declare, "I pledge to investigate and take into account the social and environmental consequences of any job opportunity I consider." The pledge caught on at other institutions, and in 1996, Manchester College of North Manchester, Indiana undertook a campaign to coordinate the dissemination of the pledge worldwide.

7 ❧ In 1997, Shell Oil took a key step in that direction by resigning from the Global Climate Coalition that had waged the long disinformation campaign against the climate scientists and announcing that it would invest increasingly in solar and wind energy technologies.

By living close to your work, schools, and other needs, you can make a major contribution to reducing the weight of your footprint on the planet. Or, put it this way: many people live in leafy suburbs and take weekend trips to the city for entertainment or "culture." In a world of endangered nature, it makes sense to reverse that: live in a place where you're surrounded by cultural diversity. Let natural diversity recover. Understand and revere the interdependence of your species with others and give the others their space.

- Reduce your *materials/energy consumption*. While pulling in city boundaries and cutting down motor vehicle use will be the biggest part of this, there's also the need to cut down on the use of materials in general. The economist Paul Hawken has noted that it took 10,000 days of plant growth to produce, through geological action, the coal or oil now used for one day of energy consumption at today's rates of use. It's like spending 30 years' worth of your life savings in one day, then going into debt to spend the same amount the next day, and the next, and the next. Yet, even so, the primary problem is not the speed with which we're using up our resources (oil reserves might last another century), but the violence their sudden release is doing to natural balances. The Earth has not evolved means of recycling wastes back into its processes this fast—whether those wastes be gasoline fumes, smoke from forest fires, sewerage, industrial effluents, or all the stuff you haul out of your home each time you move. Significantly, the biophysical stresses to the planet are more and more visibly reflected in the stresses of human lives overburdened with too many possessions—too many things to learn how to use, get repaired when they break, get parts for, find storage space for, protect from theft, get insured, shield from electric surges or computer viruses or termites, keep clean, keep out of the reach of children. Where is there time to become wise?

- Reconsider your consumption of meat. If you once thought this was only an animal rights issue, look again. To begin with, when you consume beef, you consume seven times your share of the world's grain-producing land. In a world where hundreds of millions of people don't have enough food to maintain normal work energy or health, 36 percent of the world's grain is going to feed livestock to serve the affluent. Yet for all their affluence, those who eat meat gain nothing over those who don't—and may lose. "The growing consumption of meat—particularly large quantities of high-fat meat, dairy products and eggs—is spurring a global epidemic of lifestyle diseases such as heart attacks, strokes, and cancers," says agricultural researcher Brian Halweil. Moreover, because they are large animals, large numbers of livestock (there are 1.8 billion sheep and goats, 1.3 billion cattle, and nearly 1 billion pigs) produce large quantities of defecation and disease, much of which is creeping into our water. In the United States, the amount of waste generated by livestock is *130 times* that produced by humans. A single hog farm, which recently began operation in Utah, will soon produce more waste than all the people in Los Angeles. As the spikes continue to rise, bringing higher vulnerabilities to ecological breakdown and personal illness, the benefits of a vegetarian or semi-vegetarian diet grow more persuasive.

- If your income rises, don't reward yourself by buying a bigger house, or a bigger car, or a bigger TV, or a boat! Driving a handsome four-wheel-drive vehicle just because it gives you a certain feeling of power or security or control, or acquiring a larger house just because it's prestigious or impressive, is truly fiddling while Rome burns. If you get the chance, reward yourself with a smaller abode and a simpler life that lets you live lighter on your feet.

The most extraordinary thing about this move to a more stable personal economy is that it's not just a series of separate substitutions of gains for drains, but a shift of the entire spectrum of experiences that make up a life. Recall the scenarios of the Ogallala, Nile Basin, and Thailand economies, in which the negative synergies of the spikes (or of the factors driving them) cause a cascading destruction—the deforestation, for example, increasing carbon gas emissions, which increases atmospheric warming, which precipitates more intense rainstorms, which lead to flooding, which is in turn made yet worse by the loss of percolation caused by the deforestation . . . and so on. But just as synergistic "feedback loops" can make bad effects worse, they can make good effects better. If you live in a compact neighborhood, for example, the effect of your reconfigured transport needs is not only to reduce your personal pressure on the carbon spike and extinction spike, but to improve your personal health and fitness, and thus your ability to stay connected to (and keep a protective eye on) the natural processes that keep you alive. You bike or walk more, breathe less car exhaust, have more time for direct relationships instead of the mediated relationships of drive-time radio; have more access to culture. You also have more time and energy to participate in the kind of community activism that helps others to move in the directions you've moved.

9

THE OFFER

IN 1997 the chairman of Monsanto Corporation, Robert Shapiro, told the editor of *Harvard Business Review* that he believes the most fundamental issue facing the world on the eve of the new millennium is "living within your energy income, not expending more energy than the sun provides. Also, not putting out waste products faster than nature makes them harmless." It seemed an unlikely utterance, coming from a CEO who has also said he expects his business to grow as fast in the next decade as the computer software industry did in the 1990s. Monsanto is part of a global industrial economy that now consumes 10,000 times its "energy income" each day, if we measure by the number of days of plant growth it took to produce the coal and oil that economy burns. Monsanto is also the producer of the largest-selling herbicide (Roundup) in the world—not something nature can benignly absorb.

Yet, Shapiro's comments were not just loose rhetoric. He went on to explain his view in terms that had become painfully familiar to ecologists, but that had virtually never been heard from the lips of mainstream industrialists. The world, he said, is a closed system with limited resources, and the growth of human population is on a collision course with those limits. "As far as I know," he said, "no demographer questions that the world population will just about double by sometime around 2030. Without radical change, the kind of world implied by those numbers is unthinkable. It's a world of mass migrations and environmental degradation on an unimaginable scale. At best, it means the preservation of a few islands of privilege and prosperity in a sea of misery and violence." Unless solutions are found quickly, he suggested, "we are headed for one of two disasters: either famine . . . or ecological catastrophe."

Many environmentalists reacted to this warning with cynicism, suspecting that Shapiro cared more about developing a global market for his company's new super-crops (along with the fertilizers and pesticides that had to be used with them), than about saving the planet. To some, his comments belonged in the same category of Orwellian disingenuousness as nuclear industry ads that bragged of clean air, or of paper industry ads that spoke reverentially of forests. After all, they recalled, it was Monsanto that had attacked and ridiculed the warnings about toxic chemicals that had been made by Rachel Carson in *Silent Spring*; and it was Monsanto that in the years since that book's publication had produced and dispersed enough PCBs to kill all the whales, porpoises, sea lions, and other marine mammals on the planet. It was Monsanto that was pushing the use of largely untested bovine growth hormone that was now implicated in both prostate cancer and breast cancer. And it was Monsanto, above all, that was most aggressively pushing the production of biotech solutions to agricultural pests that many scientists feared would prove even more devastating to life than the chemical ones.

Shapiro's blunt warning of impending trouble may have run contrary to the natural inclinations of the corporate culture to marginalize environmental concerns. But the kind of population collapse he was talking about is not something that can be dismissed (as many of his fellow CEOs might want to dismiss it) as unrealistic. The collapses of such ancient civilizations as those of the Sumerians, the Teotihuacáns, the Indus, and the Maya occurred because of just such collisions between human population growth and resource exploitation—and there are growing indications that they can occur again. In Russia, for example, total grain production has fallen disastrously. By the end of the 1998 harvest, it had fallen to 49 million tons—compared to 110 million tons just eight years earlier. And while most corporate leaders apparently have little knowledge of how such collapses occur, Shapiro can be assumed to have better awareness than most. His company is, after all, an agricultural-tech business, and the root causes of past population collapses were usually agricultural.

Significantly, Shapiro was not alone in his view—even among corporate leaders. In the same year, British Petroleum chairman John Browne announced that he was now convinced climate change is real, and that his company had better begin preparing for major changes. (The American Petroleum Institute immediately denounced BP for "leaving the church.") Royal Dutch Shell, the world's largest petroleum company at the time, decided that it agreed with Browne, and even withdrew from the Global Climate Coalition. World Bank chairman James Wolfensohn, shocked by the collapses—and signs of impending famine—he had witnessed in Russia and Indonesia, stepped up his efforts to convince his managers that environmental protection was becoming a critical global concern. Even a few national governments were beginning to shift their priorities. Germany, notably, had already begun experimenting with packaging laws requiring manufacturers to assume full life-cycle responsibility for their products, and in 1998 declared that it would unilaterally go well beyond the requirements of the stalled climate treaty in cutting carbon emissions from its industries.

Most remarkably, thousands of businesses (though they were still only a minuscule percentage of the global total) began quietly leaving the camp of the conventional business juggernaut. Amory Lovins, of the Rocky Mountain Institute, noted in late 1998 that even as US senators remained frozen in their opposition to the climate treaty,[1] "US leadership on climate protection has quietly passed from the public to the private sector," and that "many smart companies are already behaving as if the US Senate had ratified the Kyoto Protocol. They're becoming very clever at finding ways to turn climate protection into profits."

In one respect, these developments had to be enormously encouraging regardless what the motives of Shapiro or Browne might be; they showed that in some quarters, at least, the stranglehold of marginalization had been broken, and that those who warned that the world was on an unsustainable path could no longer be dismissed as doomsayers or panic-mongers. The scornful denials of global warming, so prominent in the mid-1990s, were now heard a little less often. In the case of Browne, it turned out that a major reason for his shift was that a growing number of BP *shareholders* were demanding that the company begin facing reality.

The deniers hadn't actually retreated from their agendas, however. Rather, they had slipped seamlessly from saying "it isn't happening" to saying "of course it's happening, but don't you see,

1 ❧ "Glaciers moving faster than climate negotiators," said one headline.

that's *good.*" That way, their message to policymakers remained the same as it had been for the past seven years: "There's no need for any restrictions on business as usual." In 1998, for example, a new conduit for this message appeared under the name of the "Greening Earth Society," which issued a very scientific-looking report concluding that that rising CO_2 is a good thing because "it enhances the growth of carrots and radishes." The report didn't mention that rising CO_2 also fosters the spread of malaria, or the growth of crop-killing weeds and pests.

The Greening Earth Society, it turned out, was yet another of the many fossil fuel front organizations set up to engage in the management of public perceptions—this one funded by the Western Fuels Association, a coal-industry trade group. But even the most discouraged climate scientists and environmentalists had to agree that there was something of a breakthrough in the new group's sudden acknowledgement—which means the coal industry's acknowledgement—that global warming is indeed happening. This shift, combined with the comments emanating from Monsanto, Shell, and BP, came like a shaft of sunlight bursting through a dark cloud. The global climate, it turned out, had undergone a far greater alteration in the years since the *World Scientists' Warning to Humanity* than even the scientists had anticipated, and by 1998 the change was simply too obvious to deny. Despite the efforts of the GCC to stop it, the debate began to shift—from whether we really want to risk our economy, to whether we really want to risk our world.

In some respects, however, the views of Robert Shapiro, even if taken at face value, will do little to move us to safer ground. His comments still reflect the narrow perspectives of the specialist—in his case, the apostle of high-volume monoculture productivity. In acknowledging that the looming food shortfall foreseen by

Lester Brown and Paul Ehrlich almost certainly will come to pass unless enormous increases in crop yields are achieved, Shapiro and his company have proposed a grand scheme for increasing global food output—a genetic redesigning of all humanity's basic foods. Yet, in the promotion of that Godlike undertaking, Shapiro expressed no trepidation about the risk that his plan will push the growing weight of humanity still farther out on a limb that is already in danger of breaking. Historically, increasing food production in order to accommodate increasing population in a region has usually resulted in the region's population increasing even more, until the health of the ecosystem or its inhabitants failed and the population either migrated to another region or collapsed. Monsanto's plan is not regional, however; it is global—and if the system fails, migration is not an option.

The Shapiro view—that it's time to respond aggressively to the crisis—might well be representative of leaders at any of a thousand other cutting-edge organizations (though that thousand would be dwarfed by the hundreds of thousands that remain unmoved). But while recognizing that the crisis is real, that view still treats the crisis from the narrow standpoint of a business opportunity; it is still firmly guided by the techno-optimist's faith in ever-newer technology. Shapiro wants the world to be saved, but he wants it to be saved by Monsanto, not by Indian village farmers choosing their own seeds and fertilizers, or by Pennsylvania organic farmers doing without pesticides or transgenic varieties altogether. His view doesn't address the needs of local communities for their own intelligent participation in land stewardship, and for their preservation of local biological and cultural diversity. Nor does it address the danger of a widening economic gap between the global corporate class of which he is a part, and the 2 billion poor who still manage most of the world's land. In short, the Monsanto plan would prop up the population spike a bit longer, while eroding biodiversity through the spread of vast monocultures, and further undermining the very world it proposes to rescue.

A particular fault Monsanto position is that it widens the chasm between private and public interest. As a CEO, Shapiro still serves his shareholders first. He wants to save the world, but on his company's terms and for his company's profit. That still means a system in which a few will profit hugely while the many will at best merely survive. It still assumes a system on which salvation will be found in growing the economy even larger, not in more equitably distributing the substantial wealth already being circulated by elites. It's a difference, in part, between assuming the privilege of borrowing without limit from other cultures, species, and generations without any real intention of repayment, and finding ways of living in balance with Earth's limited capacity for providing fresh water, food, and the protections of genetic variety.

The fear of the wealthy that they'll have to give up their "incentives to be successful"—their aspirations to have the intoxicating experience of consuming hundreds or thousands of times their own share—is a large part of what drives the campaigns to promote the grow-the-economy view and to brush off any thought of more equitable distribution. For decades, redistribution was mainly an American fear, but with the fall of Soviet communism and the rise of the global free market, it has spread to the elites—and politicians—of every center of power from Moscow to Nairobi. It's taken as an unquestioned given, now, that the way to bring developing countries out of their poverty is to grow their economies—meaning, of course, to grow the sales and exports of their biggest industries.

That fear of redistribution is no doubt reinforced by the realization that social hand out programs—whether welfare in the United States or foreign aid from wealthy nations to poor—were failures that made fools of the givers. Yet, the grow-the-economy alternative has turned out to be a pyramid scheme. Apologists may point to UN statistics showing that incomes have risen, but those incomes are based on GNP, which means only on gross totals, in which very high income at the top of the pyramid can hide crushing poverty at the bottom. When export revenue is pumped up by selling off timber, minerals, or shrimp, as has happened in a hundred countries, it often means millions of marginalized people see their incomes fall still further—yet official per capita income can still be reported as increasing, because the ballooning export income is averaged in. It's what happened in the Thai fishing villages where the shrimp mafia put local wild-shrimp fishermen out of business while adding impressively to the country's export revenue and per capita income. It's what has happened in Mexico, where hundreds of thousands of subsistence farmers have had to give up their land to wealthy cattle ranchers who expand their fortunes by selling beef to US fast food chains. And it's what has happened over the past quarter-century in India, when GNP grew impressively as a result of high volume exports of rice, iron, and cotton, yet the disparities between rich and poor only became wider than ever. In real standards of living, the world has split in two, and there are now more people at the edge of the abyss—or being elbowed into it—than ever before.[2]

2 ❧ In 1995, the World Bank made an extraordinary admission: that under political pressure from friends of the Suharto dictatorship in Indonesia, it had grossly understated the number of people living in poverty in that country after three decades of aggressive grow-the-economy policies.

In the ways that really count, the wealthy have abandoned the poor, and in doing so they've embraced the doctrine that they should abandon the idea of a common good (which has vaguely socialistic or communistic connotations), and in doing that, they've abandoned the system of tax-based public enterprise—the sharing of skills and cultivation of diversity—that is, as noted, the real basis of civilization itself. As power has shifted increasingly from public to private (private police, private armies, private prisons, private communities, and of course private business), unaccountability—and abandonment of public stewardship—has become pervasive. "Privatization is always accompanied by an upswing of corruption, all over the world," observed economist Ildiko Ekes of the Hungarian Research Institute for Economic and Social Affairs during the aftermath of the Soviet collapse.

It's not ideologically insignificant that the wealthy—first in the United States, but increasingly in Europe and Asia now as well—have in many respects abandoned the urban centers that were the seats of civilization. It may seem outlandish to suggest that the flight to wooded estates and green lawns has been anticivilized. But in fact it has been party to an ecological catastrophe—and as earlier discussion should make clear, ecological stability is an absolute requirement of a stable society. The dispersal of people over the countryside hits the ecology on a range of fronts simultaneously: it hugely increases energy consumption, because people drive cars long distances not only to commute but to run errands and ferry their kids. That drives up pollution and rogue carbon dioxide. The sprawl bulldozes habitat, which combines with the pollution from the car to undermine biodiversity—and so on. The green lawns and nursery-provided shrubs (produced by the millions in monocultural mass production) contribute to further bio-simplification. Biologically and culturally, diversity is left behind and all life becomes more homogenous.

Some of the wealthier exiles from urban society even enter a kind of garrison state, in which even small deviations from their way of life—houses that don't meet the narrow architectural covenants of the property owners' association, or that are surrounded by wild meadows instead of neatly mowed lawns, or that are shared by groups of unrelated adults instead of traditional nuclear families—are frowned on or kept out. Gated communities are proliferating in selected leafy environs all over the world, even as squatter communities and anarchic inner-city ghettos are expanding all around them.

But as suggested earlier, such retreat no longer offers protection from the real threats to human security. Climate change, disease, bioinvasions, and pollution do not stop at gates or walls. The effects of water scarcities in China or crop failures in India not only don't stop at those countries' borders but reach right into the stock portfolios or insurance policies of the people behind the gates. Even physically, the suburban enclaves are not the protection their residents once expected. It's a common lament, in exurban and small town newspapers, that crime has crept outward. But crime is only an outward sign. What's really spreading—wherever the streets and cars will take it—is the growing wave of human population, with all its increasingly urgent and consumerist wants. Many of the gatehouses will eventually be abandoned, as economic immigrants, political refugees, the jobless, the hungry, the orphaned, the lost or confused, and those mentally ill who are no longer cared for by society, come milling through.

It's not hard to sympathize with the conscious motives that lead the affluent to the suburbs, of course. One of the stronger motives may be an enjoyment of nature, and an instinctive desire to get closer to it.[3] Yet the capacity to see a bucolic quarter-acre as a piece of nature that you can possess is a sad reflection of how fragmented our vision—as well as our environment—has become. That retreat to the green quarter-acre, in a world that is going into extreme biological crisis, is a bit like cutting off your hand and putting it in a steel safe so it won't get the cancer spreading down your arm. In the long run, and now more and more in the short run too, the only way to satisfy the real needs the suburban home was intended to satisfy is to find a solution that also satisfies the needs of the squatter city or inner city.

To put this need into full perspective, though, requires another long view back. The suburbanization of human settlement is revealing a precivilized bent, not only in its hostility to tax-based investments in public life, but in its growing embrace of a world in which a person simply takes what he wants and then defends himself against any would-be competitors, human or otherwise. The wealthy individualist holed up in his trophy house is a high-tech hunter-gatherer: he may be a corporate raider, or mining executive or securities dealer, but his stance is a throwback to the preagricultural family whose objective was to hunt down and take what they needed. They were not yet ready to take the risks of engaging in the kind of cooperative enterprise—or stewardship of a commons—that requires a willingness to give, share, and trust (even while knowing that trust will sometimes be broken or tested) in a much larger enterprise than one can personally

3 ❧ Ironically, it is the wealthy who are the most enthusiastic patrons of botanical gardens, national parks, native plant societies, bird sanctuaries, and other vestiges of the natural world, even as their disproportionate consumption—and mine, and maybe yours—helps destroy it.

dominate or control. The owners of gated estates may be only a small minority of suburban dwellers, but they set the rules of the game.[4]

To cope with the changes now sweeping the world will require reconnecting with geography. That means real, physical and biological geography, not the ecologically ignorant overlays of a political map, or a real-estate dealer's map of home sale prices. For many of us, this will mean taking a hard look at the ground under our feet for the first time. And, it will mean looking at this ground with new kinds of questions not only about where to live, how to live, and what kind of work to do. It will also mean knowing—now that we see clearly the big picture kept from us before—that what is good for us as individuals is not in conflict with what's good for the local or global publics. One thing that will not work, whether in choosing a place, a life-style, or a calling, is to try to break off and go it alone.

What that means is something as revolutionary as any change in our evolutionary past—perhaps as in the history of life altogether. It means, necessarily, that in addition to pursuing a better scientific understanding of the connectedness of life, we are on the threshold of a radically different consciousness of our roles, in which every thought a person has about his or her own needs is also a thought about the needs of other life. This won't be an abrupt change; we started toward it with the institutions of

4 ❧ "The upscaling of the American dream started in the 1980s, prompted by the escalating life styles of the most affluent," says Juliet Schor. "Between 1979 and 1989 the top 1 percent of households in the United States increased their incomes from an average of $280,000 a year to $525,000. This is the group that is now widely watched and emulated, whose visible consumption is the life-style to which most Americans aspire."

parenting which we share with many other highly evolved fauna (bonding to offspring rather than just hatching them), and the institutions of marriage, family, and friendship. We've probed the connectedness through larger institutions of community, church, and nation, and in these larger connections we've undergone a vast learning experience that might be described as an evolutionary experimentation with the organization of life in various levels beyond the individual organism that is each of us. Like the evolution of species in the natural world, the process is endlessly experimental. Perhaps our organization of society has subconsciously replicated what happens in the natural world, where most of those experiments ultimately fail, but along the way produce magnificent moments. If we recognize the institutional history of human life as a part of evolution, rather than simply as a series of events in the life of a static species, it may help to explain why our history is so replete with triumph *and* tragedy, pleasure *and* pain. In real life, the failed experiments that are all in an eon's work for evolution may be deeply disappointing for sentient individuals who are a part of them.

As evolution proceeds, however, it's clear that the process was never random. Great movements were created: the movement of life from sea to land; the evident transmogrification of dinosaurs to birds; the rise of mammals. Other broad patterns we've seen coming more quickly, just in the past human generation; they have included the rising resistance of pests to pesticides; the mixing of ecosystems; and the spike of extinctions. These, among other signs, suggest we're getting closer to a time when we'll either have to end our artificial separation from the rest of life or end our span, as most other species have eventually ended theirs.

Instinctively, I think, we know we're close to that decision point—or call it an action point, since mere decisions won't be fast enough. On some level of understanding that is organic and not yet altered beyond reach by ideological and commercial conditioning, we know we've become separated from the processes that sustain us, and we ache for explanation: it accounts, perhaps, for why a specter of apocalypse can be found in so many of Earth's religions and cultures.

The evolutionary change we started toward with the institutions of parenting, marriage, and family, and to which I think we have to complete the transition if our species is to remain a part of what might be called God's plan, is to develop a state of consciousness in which the sense of identity each of us feels is not housed only in our individual bodies and minds, but in the ecological processes of which these bodies and minds are a part. Some belief systems have touched on this, but the idea has remained marginalized by the focus of media on celebrity, which is the product of a heightened fascination with the mystique of the individual.

But there are signs that the border between the individual as a distinct entity and the sociobiological world around him may be dissolving much in the way national borders as artificial envelopes of identity are. That doesn't mean the time is approaching when we won't be able to look in a mirror and see ourselves as recognizable individuals. But perceptions are shifting, as we become increasingly conscious that nothing in ourselves is permanently owned. The air you inhaled this morning may have blown across a distant mountain range last night; the protein that is rebuilding the leg muscles you stressed running for a plane last night may have been part of an anemia-infected North Atlantic salmon last week; the water you perspire this afternoon may have been a part of the anoxic Black Sea last year and the melting Antarctic ice cap the year before and the sap of an adelgid-diseased hemlock the year before that. As we become more aware of ourselves as processes rather than objects, our infatuation with ourselves as objects—as the consumers who constitute an enormous global market for cosmetics, for example—may itself begin to dissolve. The idea of such dissolution may be extremely frightening, since it may be taken to mean a loss of self, or a loss of consciousness. However, it may actually entail an *expansion* of consciousness, and of the satisfactions of love and life that that expansion makes possible.

The deepening footprint of humanity is starting to feel too much like quicksand. Yet despite our collective confusion and still-widespread obliviousness, it would be far from the truth to say no one is moving hard to get us out. Tens of thousands of analysts, organizers, educators, and activists are at work on thousands of projects to stabilize population, cut carbon gas emissions, save biodiversity, and bring consumption under control. But their numbers are to the numbers of the oblivious as the rich are to the poor: they are still a tiny minority, and the gap may be widening.

The concern now is not whether we have the technology and the intelligence to continue; it's about whether we're putting it all together fast enough. If we're on track to complete the building of an ark in a month, but the flood will be here in an hour, the ark won't help.

Consider, for example, the technology of energy for electricity. In the 1990s, solar technology began to boom. By 1998, over half a million homes in the developing world were getting electricity from photovoltaic solar panels. This is great news, except for one thing: the solar boom still only accounts for one of very 4,000 watts of the world's total installed electric capacity. Obviously, it's not good enough. The boom will have to redouble and redouble and redouble—year after year. And of course we know that's possible, because that's how all successful new technologies begin. When the Model-T car began to compete with horse-drawn carriages, the same kind of daunting prospect lay ahead. But for every Model-T that makes history, there are thousands of inventions, dreams, or experiments that fail or are suppressed. Ideas are like acorns: in an acre of forest, thousands of them may be scattered about, from which scores of saplings may start up and grow fast, but only one may survive to become a great oak.

Our challenge is to put ourselves in the position of that one acorn. And, though the odds seem small, it's well to recall that mathematically, it's already something of a miracle that we're here. One of the most significant ways in which science and religion converge is in the recognition that miracles are not random. Whether they occur because of natural laws or divine plans, most of those who believe that we humans bear some responsibility for our fates would agree that momentous choices lie ahead for us. If it took a miracle to get us where we are, it will take another to move us forward on our evolutionary path. That path has become extremely treacherous, and we're going to have to pick our way with great care.

It will take not just a great multiplication of the small efforts that have put solar roofs on huts in Nigeria, or treadle pumps (giving peasant farmers access to groundwater) on fields in Bangladesh; it will take a coalescing of these efforts so that instead of viewing them as admirable but isolated projects, we begin to see them as part of a new, transcendent pattern that affects all of our thinking. We need to reform our vision so that we are able lead the way for our descendents, not blunder forward with our heads turned to fight a rear-guard battle against our ancestors, most of whom may have been completely unaware of where they were leading us.

By recognizing what has blocked our vision until now, we can identify some general rules for forming the kinds of policies and practices that will make this miracle happen.

See the scale of things. The booms in solar and wind energy are seeds that have sprung up in a fertile soil. But they could be killed easily. Every conscious human decision involving energy and its uses—whether in transportation, urban design, or resource extraction—needs to take into account this ultimate goal of nurturing the seedlings of the major industries of the future. A climate treaty that promises to reduce CO_2 emissions to 5 percent below 1990 levels by 2010 will be worthless if it is not quickly revised to 60 or 80 percent. Laws that save a few hundred endangered species will do little to save the planet's life as a whole, if not supported by changes in human culture that can turn those few hundred rescues to millions. "Without perspective, we are lost," says writer Tom Athanasiou. "Good news, most of it from the United States and other rich regions, does not automatically scale to the planetary level."

Look at the connections, as closely as at the things connected. In our view of the world, conditioned by centuries of habit, even though we know things are in constant motion, we tend to think it is the things that are the reality, rather than the motion. For example, we own cars, but we don't think of owning (or assuming responsibility for) the motion of the car, or the processes (consumption, combustion, emission, invasion) the car is a part of. This traditional view comes from the fact that things are visible; processes are not. But as we learn more about the ecology of the earth, we begin to understand that the processes are as real, as fragile, as vulnerable to damage or distortion, as the palpable things. In fact, some scientists take this observation further and point out that on a subcellular or subatomic level, even the things are made up only of processes. But this rule isn't just theoretical; it has important implications for action. Among the most important: in education and research, we need to make connections our main focus. Children in elementary science, for example, need to be taught about water not simply in terms of its chemical makeup and physical behavior, but in terms of the hydrological cycle that could make or break the human future.

Take into account other communities, cultures, generations, and species. Just as it would be foolish to allow long-term planning to rest entirely in the hands of particular agencies or administrations of government, it would be foolhardy to ignore our relationships with the other dimensions of life—whether geographical, temporal, or genetic. If we kill them, we kill us. As environmental scientist Jesse H. Ausubel advises, "We must take seriously the Copernican insight about Earth's position in the cosmos and not simply replace geocentricism with anthropocentricism."

Know the sources of our information and our beliefs and do reality checks on how much we rely on mediated ideas. We shouldn't be afraid to use technologies like computers or TV as multipliers of our powers, but should be wary of whose powers are being multiplied, and to what purpose. We need to find ways of preventing the control of public beliefs from falling into private hands, whether for rogue ideological purposes or for concentration of wealth. Ultimately, this requires a clear separation of the funding for science and education from the largess of industry. From McDonalds franchises in public schools to Genentech grants to the university biology research programs to Mobil Oil supporting public radio reporting on climate change, it's a slippery slide to just sit back and let our knowledge be packaged and paid for. In the long run, our survival depends on our insistence that the findings of science—and the accumulated knowledge that is the legacy of real civilization—both be paid for by public taxes and remain accessible to all humanity. Meanwhile, we need always to keep in touch with core sources of primary information that we know to be trustworthy. To be trustworthy is not necessarily to be right, of course; our parents can be (and often are) wrong, and our senses can fool us. But those sources are not systematically aligned to exploit or manipulate us.

Beware of living too heavily in a world of fictional experience and entertainment. Bewarc of drifting into heavy reliance, for stimulation and companionship in the adventure of life, on technologically mediated surrogate parents, friends, companions, or elders. As communities and as individuals, if we lose touch with direct experience and direct contact—both with other people we trust and with a physical environment we understand and trust—we will lose our lives.

Observe the precautionary principle in all uses of technology. This principle says, simply, that in assessing the risks of any technological expansion of our powers, the burden of proof of safety must be on those who promote the expansion, not on those who oppose it, as is normally the case now. Short of scientific and public consensus that the risks are acceptable, the proposed use should bc forbidden. To a dangerous degree, the genie has long been out of the bottle with this principle (chemical manufacturers can release a new compound into the marketplace as long as no one has proven it *isn't* safe), but that makes it all the more critical that we put the cork back in. We've seen—and come to universally regret—the mistakes we made with nuclear weapons proliferation. We're beginning to see the magnitude of the mistake we made in allowing more than 100,000 synthetic chemicals to be produced and dispersed throughout our air, water, and soil, when the long-term effects on people and the planet have never been tested. In effect, we have allowed our industries to use us and our world as their guinea pigs—with no recourse if things go wrong. And now we're starting down that same treacherous path with biotechnology, in a reckless attempt to extend our agricultural capacity—and our population-carrying capacity—still farther out on a limb that is already breaking. It's time to reverse the practice of allowing transnational companies to patent and profit first, then worry (or not worry) about what the new chemicals or organisms will do to the planet's life.

Look beyond technology. In the late twentieth century, technology came to be implicitly treated as the ultimate human achievement. (Yes, there are those who have warned us that it is not, but they have been easily marginalized as "Luddites"—or "extremists who want us to go back to the Stone Age.") Yet, a technology is only a tool, an extension of the capabilities of our hands, eyes, and ears. There are other kinds of tools that extend the reach of our minds and spirits. We have accelerated the development of those tools with the computer revolution, but it would be a mistake to regard what has happened on that front so far as anything more than a beginning. Computers today still extend mainly the brain's more mechanical functions: calculation, organization, storage, and transmission of binary bits in much the same way that trucks and rail cars carry the grains of sand that are used to make concrete. The higher faculties—the tools that succeed technology—still lie ahead. Again, there are important practical implications. It is a colossal foolishness, for example, to assume that what we need in order to fix our broken educational system is to put computers in all the schools. That's an extension of the same doctrine that was espoused by World Bank developers who decided decades ago that what would liberate the Third World was a proliferation of big power plants, highways, and office buildings. Many of those projects left only greater poverty. What children need now is not more extension of their already vastly extended corporeal powers, but more capability to make sense of the powers they have.

Begin complete materials and energy accounting in all human activities. From original extraction (mining, logging, oil drilling, water harvesting) to industrial processing and manufacturing to use of products and services to final disposal and recycling, we need to establish a worldwide system for tracking all materials and fuels from nature through the economy and back to nature. The system needs to be complete, with no gaps for corporate or national secrecy, or shadow activity, or careless or surreptitious dumping. The system also needs to be protected from policies that would restrict scientists' access to information that is part of the common heritage of humanity. This database will be essential to the effective functioning of the new forms of governance we will be needing in the coming century.

Move to complete financial accounting, in parallel with the tracking of physical and biological materials and of human labor. In the same way that not having to pay for pollution (as with CO_2 emission "allowances") weakens and destabilizes the environment biologically, not having to pay equitable wages for labor (as with garment sweatshops) weakens and destabilizes society. An unpaid dumping of waste, or a low-paid exploitation of labor, is still a cost. At every stage where such costs are incurred, prices need to reflect them. In this way, hidden costs can be paid by those who are responsible, rather than passed off to those who are not.

Move toward global information tracking, in parallel with

materials/energy and financial accounting. Treat the integrity of public information with the same rigor as we treat the safety of food or drugs, since systematic misuse of information can do the same damage as deliberate poisoning or addicting. We need to establish the same kinds of standards for science reporting, advertising, and corporate PR as we do for publishing medical information. They have no less potential (and often more) to affect the lives of millions.

End the private ownership of public policy. Environmental economists make a strong case that in places like India, Mexico, and Indonesia, where assets such as forests and water were once commonly owned, the appropriation by states and subsequent "sale" to private interests has destroyed traditional stability. Private ownership, they argue, should be reformed in two key ways: the price of ownership of anything physical or biological should include full costs of production and recycling to full future use; and certain assets that are inherently common assets of all the world's life—not just its people—should not be privately owned at all. Meanwhile, however, as we begin to understand more clearly that the assets are not just physical materials but the processes of which those materials are a part, the owners of the property have made rapid headway in acquiring effective ownership of the processes as well. Ultimately, this is the greater danger. We see it in phenomena like the ability of US politicians to keep in effect a nineteenth-century mining law that allows the corporations that finance them to take wealth from Earth (often causing heavy pollution and illness to nearby residents in the process) virtually free. The mining law thus belongs to the mining companies, who hold on to it by keeping it out of public view. The same process can be extended—and is being extended—to laws affecting a wide range of industrial processes and communications services that affect what we know and believe. By controlling the information sources that shape policies, it is possible to control the human interventions that shape nature. To do that is to grab the reins from God without knowing where the horse is headed. Some call this reckless. Others call it blasphemous.

Shift from the economics of liquidation to the economics of restoration and asset building. As destructive industries like fossil fuel extraction, subsidized factory fishing, and industrial timbering are phased down, new industries aimed at restoring assets, stabilizing climate, and protecting biological diversity will take their places. In the future, we can anticipate that a major new industry will be needed to clean Earth's soil, water, and air, all of which have functioned as filters or sinks for too many pollutants, too long. New solar and windpower industries, and eventually a major new hydrogen-energy industry, will take over from oil and coal. Other industries will arise around the close monitoring of natural processes and human activities that spring from them. Just as the human economy will need to better replicate the economy of nature to be sustainable, the information systems it uses for self-regulation will need to better replicate those that operate in nature—from the nervous and endocrine systems of individual organisms to the genetic adaptations of species. In many cases, it may be the obsolescent industries of today that reform their missions and methods to *become* the new industries of tomorrow.

Reconsider "growth." In this, we come full circle, to seeing the scale of things. Just as sleeping pills or vitamin A can be beneficial in one quantity and lethal in another, so it is with all forms of growth. Rampant growth in cells becomes a cancer; in algae it can become a lake-killing "bloom"; in a population it can lead to implosion. In the human economy, an industrial metabolism that sucks up resources faster than they can be reintegrated into the natural environment can poison itself. It is not just unrestrained growth, but the ideological promotion of growth that has driven the spikes on which our civilization has been impaled.

Come back to the theme of the pervasive denial that lets so many of us believe we're not really responsible for the consequences of our choices—that God is, or that "human ingenuity in the future" is. The culture of high consumption, along with the conveniences of high technology, have made us accustomed to the idea that somehow we will always be provided for. But that kind of passivity is a sign that we're in danger of losing the kind of instinct for survival—what author Jared Diamond calls the "cunning"—that has enabled us to make our epic journey across the millennia. If we think we'll keep on making it because we're chosen, we're becoming lulled by hubris; we're losing our edge. It's possible that the farmers in India and China and France, rather than be guaranteed success by the central management of Monsanto, would prefer—or even need—to choose which variety of seed to plant themselves. That need may be as strong as their need to eat.

The vast majority of the species that have lived on this planet have run their courses and given way to successors, and we live by the same biological and evolutionary rules as they. "The evolutionary unity of humans with all other organisms is the cardinal message of Darwin's revolution for nature's most arrogant species," says biologist Stephen Jay Gould. The idea that God will provide can become a dangerous seduction. Only 2 percent of the species that have ever lived on Earth live on it now, and that percentage is falling as the spike of extinctions rises. Most of the other 98 percent came and went before our species emerged. Extinction, when it occurs at a natural rate, has always been as normal a part of evolution as has the formation of new species. If we can draw a generalization from several hundred million years of experience, it is that every species, sooner or later, either adapts to changing conditions or dies out. There is no reason to believe—it would be fatally arrogant to believe—that the human species is exempt from this rule. On the contrary, by hastening the demise of thousands of other species on which our food, water, and health depend, we could usher ourselves out along with them.

The deniers—those who say that global warming is no problem, or that extinctions of obscure birds or reptiles don't matter—may be less visible now than they were a few years back, but they still command superior numbers. The Business Roundtable, the Global Climate Coalition, and the apostles of high growth have not gone away. If people like Al Gore, Robert Shapiro, and James Wolfensohn are to make good on their claims of leadership toward a sustainable society, they'll need to shrug off the blinders of political expediency and take some large personal risks. Shapiro will have to risk losing his job, Gore losing an election, Wolfensohn forfeiting some of his US funding.

What will such leadership require? It will mean having the courage to sweep away the foolishness of the climate treaty's minuscule CO_2 reductions and to go for a full phaseout of coal- and oil-fired energy systems—and their replacement by solar, wind, and hydrogen energy production—in the next 20 years. It will mean shaking off the intimidating tactics of anti-abortion zealots and instituting a full-scale, worldwide family-planning program to stabilize population the humane way—before AIDS, toxic pollution, famine, or civil riots and disruptions do it the hard way. It will mean redesigning all our industries and products so that the services we need are provided equitably worldwide, yet with only a fraction of the material resources used now. It will mean pulling in the boundaries of our cities and towns both to reduce the consumption of transportation energy and land and to let nature regenerate. It will mean restoring the natural boundaries of biologically distinct regions, even as the boundaries of obsolescent nations continue to dissolve. And, it will mean overhauling our systems of economic incentives to reward behavior that protects rather than exploits natural resources and that penalizes the heedless borrowing—or theft—of what is needed by other generations, cultures, or species.

Since all these actions are likely to be labeled extremist, it's pragmatic to ask why anyone would take such risks. One answer is that there's no longer any possibility of avoiding a large risk one way or another. Either you take risks personally, or you take them as a shared enterprise with the rest of humanity. Let's consider the choices.

On one hand, there are the kinds of personal risks with which all people have been preoccupied since the dawn of history: of physical attacks on you or your loved ones; of life-threatening hunger or thirst; of infectious disease; of unemployment; of loss of liberty or violation of personal rights. Over the millennia, we've established institutions to protect the integrity of the individual—first through the clan or tribe, then the nation, now UNDP or WHO or Human Rights Watch. Few would deny that this integrity is important—that it is, for most of us, our very identity.

On the other hand, there are the risks imposed by the unprecedented spikes of population expansion, carbon dioxide concentration, consumption, and extinction. The *World Scientists' Warning to Humanity* gave us fair notice that our unwitting alterations of the air, water, and land are subjecting us—all of us, without exception—to an enormous experiment after which no second chance will be possible. Biologists, for their part, make the more specific warning about mass extinctions: that if the experiment fails and it turns out that we have destroyed too much of the web of life to keep its basic processes intact, we could be inviting a precipitous crash that will take us down too.

These global risks are less familiar to us than the personal ones, and we have far less experience in preparing for them. Most of us know what we have to do to keep our jobs, or protect our homes from fire, or treat an infection. We have elaborate mechanisms in place to address those risks. But few of us have thought about what to do if our biosphere should begin to fail. Few institutional responses are in place. Even an agency like the US Federal Emergency Management Administration (FEMA) still operates under the obsolescent idea that large weather disasters are anomalies. So our mind-set is unprepared for what is upon us. Our impulse is still to guard the personal stakes first. Yet, a personal insurance policy is no good if the insurance company has been driven out of business. A national government's rescue of failing businesses can't succeed if the government's tax base has been swept away. The tax base can't survive if the resource base has failed.

In effect, we have been presented with an offer. We can continue to focus on guarding ourselves individually, in our gated communities or guarded nations, as long as possible—but well knowing that what we're doing can't go on much longer. Or we can turn our attention to something far more enduring.

The offer is to trade our closely guarded personal security for the larger security of the world we stand on. It need not be a final sacrifice of selfhood, but merely a brave step that will in evolutionary time take only a moment, but that will make—as Robert Frost wrote—"all the difference." Like the anonymous bystander who once leaped unhesitatingly into the ice-cold Potomac River to save drowning passengers of a crashed airplane, we need to cast off whatever fear has held us back until now. Ironically, it can be only through our acceptance of this offer—to defend our world instead of ourselves—that we have any real chance of saving ourselves and of regaining the sense of personal and family security we care about so deeply.

Do it now and our leaders will follow.

NOTES

INTRODUCTION

(NOTES FOR PAGES 1–8)

Collapse of the Russian Grain Harvest

John-Thor Dahlburg, "Hunger in Russia's Heartland," *Los Angeles Times*, October 20, 1998; US Department of Agriculture.

Fires

The World Wide Fund for Nature (WWF) reported that in 1997, more tropical forest burned in the world than in any previous year recorded. In 1998, the number of wildfires increased.

Global Warming

"January-July Global Surface Mean Temperature Anomalies," National Climatic Data Center of the National Oceanic & Atmospheric Administration (NOAA), August 1998.

Flooding

"Floods Wreak Havoc Across Asia," CNN Interactive, August 27, 1998; "Flooding Rampant Worldwide," CNN Interactive, September 10, 1998.

New Outbreaks of Disease

Anne Platt, "The Ecology of Infection: What Ebola, AIDS, TB, and

Other Outbreaks Have in Common," *World Watch*, July/August 1995; David L. Heymann, MD, editor, "The New Plagues," *Red Cross Red Crescent*, issue 3, 1998.

Assaults No Country Can Defend Against

US Secretary of Defense William Cohen, *Quadrennial Defense Review*, 1998; Cohen remarks on "transnational operations that defy traditional means of influence," C-Span radio in September 1998.

Citizens of No Country

Vernon Loeb, "A Global, Pan-Islamic Network," *The Washington Post*, August 23, 1998; Andreas Ruesch, "Talibanistan—Afghanistan's Mysterious Government," *Swiss Review of World Affairs*, May 1997.

Forcing Children into Battle

During a 2-year span in the late 1990s about 250,000 children between the ages of 7 and 17 years old were forced into combat in 33 armed conflicts in 26 countries, according to the Swedish Save the Children Fund, reported in "Throwing Children Into Battle," *World Watch*, May/June 1997; Center for Defense Information, *Child Combatants: The Road to Recovery* (documentary film), Washington Coalition on Child Soldiers, 1998.

Stories of Great Floods

David Suzuki and Peter Knudtson, *Wisdom of the Elders: Honoring Sacred Native Visions of Nature* (New York: Bantam Books, 1992).

Aborigines' Non-Reaction to the Appearance of a Great Sailing Ship

Robert Hughes, *The Fatal Shore: A History of the Transportation of Convicts to Australia, 1787-1868* (London: Folio Society 1998). Originally published in 1986.

20 Miles per Second

The voyage of the US spaceship *Mariner 10* covered 250 million miles in 21 weeks, as described by Kenneth F. Weaver in "Flight to Venus and Mercury," *National Geographic*, June 1975.

THE FOUR SPIKES

(NOTES FOR GRAPHS, PAGES 17, 27, 39, 41)

The Carbon Dioxide Spike

Data are from Mauna Loa measurement site, Hawaii and ice core measurements, Antarctica, reported in Intergovernmental Panel on Climate Change (IPCC), *Climate Change 1995: The Science of Climate Change, Summary for Policymakers and Technical Summary* (New York: United Nations Environment Programme and World Meteorological Organization, 1995); Seth Dunn, "Carbon Emissions Resume Rise" in *Vital Signs 1998* (New York: W.W. Norton, 1998).

Extinction Spike

Edward O. Wilson, *The Diversity of Life* (Cambridge, Massachusetts: Harvard University Press, 1992); United Nations Enviroment Programme, *Global Biodiversity Assessment* (UK: Cambridge University Press, 1995); The World Conservation Union (IUCN), the Smithsonian Institution, the World Wildlife Fund, the Nature Conservancy, the Royal Botanic Gardens of the United Kingdom, et al., *IUCN Red List of Threatened Plants, 1998*; "National Survey Reveals Biodiversity Crisis—Scientific Experts Believe We Are in Midst of Fastest Mass Extinction in Earth's History" (news release), American Museum of Natural History, New York, April 20, 1998.

Consumption Spike

W. W. Rostow, *The World Economy: History and Prospect* (London, 1978); Eric Hobsbaum, *The Age of Empire, 1875-1914* (New

York: Vintage Books, 1989); Alan Thein Durning, *How Much Is Enough? The Consumer Society and the Future of the Earth* (W. W. Norton: New York, 1992); Lester R. Brown, et al., *State of the World 1999* (New York: W. W. Norton, 1999).

Population Spike

Clive Ponting, *A Green History of the World: The Environment and the Collapse of Great Civilizations* (London and New York: Penguin, 1991); United Nations Department of Economic and Social Affairs, Population Division, *World Population Prospects: The 1998 Revision* (New York: 1998); Brian Halweil, "Curbing Population: Without Family Planning It Isn't Going to Happen," *Christian Science Monitor*, October 21, 1998; Lester R. Brown, Gary Gardner, and Brian Halweil, *Beyond Malthus: Sixteen Dimensions of the Population Problem*, Worldwatch Paper 143 (Worldwatch Institute: Washington, DC, 1998).

CHAPTER 1

(NOTES FOR PAGES 9–45)

Climate Scientists' Warning

Intergovernmental Panel on Climate Change (IPCC), *Climate Change 1995: The Science of Climate Change, Summary for Policymakers and Technical Summary* (New York: United Nations Environment Programme and World Meteorological Organization, 1995).

Hurricanes and Atom Bombs

Lyall Watson, *Heaven's Breath: A Natural History of the Wind* (London: Hodder & Stoughton Ltd., 1984)

Hurricane Mitch

"Mitch Forces Storm Rethink," BBC News Online Network, November 14, 1998.

"Report From Central America: The Devastation of Hurricane Mitch," *Ouroboros* (newsletter of the Goldman Environmental Prize), December 1998.

Weather Damages

Christopher Flavin, "Pre-Buenos Aires Climate Briefing for Decision Makers and the News Media" (news conference), Worldwatch Institute, Washington, D.C., October 20 1998; Christopher Flavin in Lester Brown et al., *Vital Signs 1998* (Washington, DC: Worldwatch Institute, 1998).

Rate of Increase in CO_2

Long-term increase from Antarctic ice core measurements reported by Robert T. Watson et al., "Greenhouse Gases and Aerosols" in *Climate Change, the IPCC Scientific Assessment*, eds. J.T. Houghton, G.J. Jenkins and J.J. Ephraums, CUP, 1990 (updated); Increase since 1957 from Mauna Loa Observatory, Hawaii, compiled by C. D. Keeling, Scripps Institute of Oceanography and P. Tans, National Oceanic and Atmospheric Administration of the United States (NOAA), reported in John Houghton, *Global Warming: The Complete Briefing*, Second Edition (Cambridge, UK: Cambridge University Press, 1997).

GCC's $14 Million Ad Barrage

Eco (NGO newsletter), Kyoto, Japan, December 5, 1997.

3 Tons of CO_2 Emissions Per Person

Seth Dunn, Worldwatch Institute, based on data from Oak Ridge National Laboratory, British Petroleum, and the Population Reference Bureau.

Kinza Clodumar's Warning

Seth Dunn, "Dancing Around the Key Climate Issue: Can the North and South Get in Step?" *World Watch*, November/December 1998.

"Historic step"

Christopher Flavin, "Last Tango in Buenos Aires," *World Watch*, November/December 1998.

Scientists' Warning

World Scientists' Warning to Humanity, Union of Concerned Scientists, Cambridge, Massachusetts, 1993. The warning was signed by scientists of 71 countries including all of the 19 largest economic powers and the 12 most populous nations.

The variety of organisms, once lost . . .

Edward O. Wilson, "Biophilia and the Conservation Ethic" in Stephen R. Kellert and Edward O. Wilson, *The Biophilia Hypothesis* (Washington, DC: Island Press, 1993).

Decline of Birds

Birdlife International, Cambridge, England, cited in Howard Youth, "Flying Into Trouble," *World Watch*, January/February 1994.

Decline of Mammals

IUCN-World Conservation Union, cited in Curtis Runyan, "Mammals in Global Decline," *World Watch*, January/February 1997.

Decline of Primates

John Tuxill, "Death in the Family Tree," *World Watch*, September/October 1997.

11 of 15 Major Fishing Grounds

Maurizio Perotti, Data and Statistics Unit (FIDI), Fisheries Department, FAO, Rome, cited by Anne Platt McGinn in *Rocking the Boat: Conserving Fisheries and Protecting Jobs*, Worldwatch Paper 142 (Washington, DC, Worldwatch Institute, 1998).

Return of the Potato Blight

International Potato Center, Lima, Peru, reported in *World Bank News*, June 4, 1998; Fred Pearce, "The Famine Fungus: Potatoes Could Feed a Billion People, Yet the Return of an Old Enemy Would Turn Plenty into Disaster," *New Scientist*, April 1997.

Corn, Beans, Spinach

US Department of Agriculture, cited in Cary Fowler and Pat Mooney, *Shattering: Food, Politics, and the Loss of Genetic Diversity* (Tucson: University of Arizona Press, 1990).

75 Percent of Genetic Diversity in Agriculture Lost

Hope Shand, "Bio-Meltdown," *New Internationalist*, March 1997.

Exceeding the Value of the Entire Human Economy

Robert Costanza et al., "The Value of the World's Ecosystem Services and Natural Capital," *Nature*, May 15, 1997.

Red List of Plant Extinctions

The World Conservation Union (IUCN), the Smithsonian Institution, the World Wildlife Fund, the Nature Conservancy, the Royal Botanic Gardens of the United Kingdom, et al., *IUCN Red List of Threatened Plants*, 1998.

Fastest Mass Extinction in Earth's History

"National Survey Reveals Biodiversity Crisis—Scientific Experts Believe We Are in Midst of Fastest Mass Extinction in Earth's History" (news release), American Museum of Natural History, April 20, 1998.

Consuming Genetic Resources 1,000 to 10,000 Times as Fast

Edward O. Wilson, *The Diversity of Life* (Cambridge, Massachusetts: Harvard University Press, 1992).

Australian Forest Disappearing at Rate of 60 Soccer Fields Per Hour

Ashley T. Mattoon, "Bogging Down in the Sinks," *World Watch*, November/December 1998.

Forest Loss 1 Percent Per Year

World Wide Fund for Nature, cited in "Uprooted Forests" (back cover), *World Watch*, January/February 1997.

Laminated I-Beam

David Malin Roodman and Nicholas Lenssen, *A Building Revolution: How Ecology and Health Concerns Are Transforming Construction*, Worldwatch Paper 124 (Washington, DC: Worldwatch Institute, 1995).

Efficiency of Rail Transport

Marcia D. Lowe, *Back on Track: The Global Rail Revival*, Worldwatch Paper 118 (Washington, DC: Worldwatch Institute, 1994).

Synthetic Chemicals at Large

Jennifer Mitchell, "Nowhere to Hide: The Global Spread of High-Risk Synthetic Chemicals, *World Watch*, March/April 1997.

CHAPTER 2

(NOTES FOR PAGES 47–89)

Red River Rising

Bob Moen, "North Dakota Floods Break 100-year Record; Homes Evacuated but Dikes Hold," Associated Press, April 17, 1997; "Red River Rises to Highest Level Ever, Busting through Levees, into Homes," *Detroit Free Press*, April 18, 1997; "North Dakota Struggles to Stem Worst Flood in 500 Years," Agence France Press, April 19, 1997.

A Once-in-500-Years Event

Elizabeth Kane, "Floodwaters Take Toll on Dakota Engineers," *Engineering Times*, volume 19, number 6, June 1997.

20 Million Flooded Out of Homes in Bangladesh

Celia W. Dugger, "Monsoon Hangs On, Swamping Bangladesh," *The New York Times*, September 7, 1998.

Storm Damages

Christopher Flavin, "Pre-Buenos Aires Climate Briefing for Decision Makers and the News Media" (news conference), Worldwatch Institute, Washington, DC, October 20, 1998

"Loopholes the Entire Fossil-Fuel Industry Could Pass Through"

Eco (NGO newsletter), Kyoto, Japan, December 5, 1997.

Red River Follow-up Story

William Claiborne, *The Washington Post*, April 21, 1997.

False Spectrum

"Dust in the Wind" (editorial), *The Columbus Dispatch*, August 15, 1996.

National Park Wrecked

Potomac Basin Reporter, Interstate Commission on the Potomac River Basin, Rockville, Maryland, November/December 1996.

California Dream House

John McPhee, *The Control of Nature* (New York: Farrar, Straus and Giroux, 1989.)

Valiant Stand against Nature

Debbie Howlett, "River Rising to 100-Year Flood Crest," *USA Today*, April 17, 1997.

Al Gore in Red River Valley

Dale Jamieson, "Warm Globally, Flood Locally" (commentary), *Minneapolis Star Tribune*, April 16, 1997.

"One-Man Heatwave"

Jonathan H. Adler, "Hot Air," *National Review*, August 17, 1998.

The Desolate Year

John Stauber and Sheldon Rampton, "Silencing Spring: Public Relations and Private Interests," *Earth Island Journal*, winter 1995-96.

Coal Executives' Memo

Ross Gelbspan, *The Heat is on: The High Stakes Battle over Earth's Threatened Climate* (New York: Addison-Wesley, 1997).

"Megaproblem or Not?"

Julian L. Simon, quoted in "The Global Environment: Megaproblem or Not?" (debate), *The Futurist*, March/April 1997.

110 Million People Killed in Twentieth Century

William Eckhardt, "War-Related Deaths Since 3000 BC," *Bulletin of Peace Proposals*, December 1991; Ruth Leger Sivard, *World Military and Social Expenditures 1996* (Washington, DC: World Priorities, 1996).

The Problem with FAO and World Bank Food Forecasts

Lester R. Brown, "Facing Food Scarcity," *World Watch*, November/December 1995.

Beijing Government Replies

Ed Ayres, "Note to Readers," *World Watch*, January/February 1995.

Population in the Sixteenth Century

Clive Ponting, *A Green History of the World: The Environment and the Collapse of Great Civilizations* (New York and London: Penguin Books, 1991).

Crops under Assault by New Pests

Chris Bright, *Life Out of Bounds: Bioinvasion in a Borderless World* (New York: W.W. Norton, 1998).

Pesticide Resistance in 1990

Cary Fowler and Pat Mooney, *Shattering: Food, Politics, and the Loss of Genetic Diversity* (Tucson: University of Arizona Press, 1990).

People Forced Out of Their Homes or Homelands

Kathleen Newland, "Refugees, the Rising Flood," *World Watch*, May/June 1994; United Nations High Commissioner for Refugees, 1997.

Water is Now a Life-or-Death Issue

Anne Platt, "Water-Borne Killers," *World Watch*, March/April 1996.

An Apocalyptic Collision

Timothy Ferris, "Is This the End?" ("Annals of Space"), *The New Yorker*, January 27, 1997.

Science "Unimportant" to Economists

Paul R. Ehrlich, Anne H. Ehrlich, and Gretchen C. Daily, *The Stork and the Plow: The Equity Answer to the Human Dilemma* (New York: Grosset/Putnam, 1995).

Nike Sport Shoes

John C. Ryan and Alan Thein Durning, "The Story of a Shoe,"

World Watch, March/April 1998, excerpted from *Stuff: The Secret Lives of Everyday Things* (Seattle: Northwest Environment Watch, 1997), with editor's introduction by Curtis Runyan.

Black Markets or Unregistered Work

Ed Ayres, "The Expanding Shadow Economy," *World Watch*, July/August, 1996.

Women's Work Undervalued by $11 Trillion

Human Development Report (New York, United Nations Development Programme, 1995).

CHAPTER 3

(NOTES FOR PAGES 91–123)

The Population Bomb

Paul Ehrlich, *The Population Bomb* (New York: Ballantine Books, 1968).

Investor Worries about Surprise Attacks

Aymo Brunetti and Beatrice Weder, *Investment and Institutional Uncertainty: A Comparative Study of Different Uncertainty Measures*, International Finance Corporation Technical Paper Number 4, The World Bank, 1997.

World Bank Hypocrisy on Power Plant Loans (footnote)

Christopher Flavin, "Banking against Warming," *World Watch*, November/December 1997.

Flow of Private Capital

Jack D. Glen and Mariusz A. Sumlinski, *Trends in Private Investment in Developing Countries*, IFC Discussion Paper 34, The World Bank and International Finance Corporation, 1998.

Decline in Nuclear Plant Construction

Nicholas Lenssen in Lester R. Brown, et al., *Vital Signs 1998* (New York: W.W. Norton, 1998).

Ogallala Aquifer Depletion

Sandra Postel, *Last Oasis: Facing Water Scarcity* (New York: W.W. Norton, 1997).

People Will Die as You Read This

Paul Hawken, *The Ecology of Comm*erce (New York: Harper-Collins, 1993).

Egypt Becoming a Net Grain Importer

Brian Halweil, Worldwatch Institute, Washington, DC, 1998, unpublished.

Population Grows Geometrically

Thomas Robert Malthus, *An Essay on the Principle of Population*, 1798.

Ecosystems as Macro-organisms (footnote)

Stephen H. Schneider and Penelope J. Boston, editors, *Scientists on Gaia* (Cambridge, Massachusetts: The MIT Press, 1991).

Village under the Sea

Sandra Postel,"Where Have All the Rivers Gone?" in *The World Watch Reader on Global Environmental Issues*, Lester R. Brown and Ed Ayres, editors (New York: W.W. Norton, 1998).

Population Projections

World Population Prospects: The 1996 Revision, United Nations Department of Economic and Social Affairs, Population Division, October 1996.

Famines

Daniel Leviton, editor, *Horrendous Death, Health, and Well-Being* (New York: Hemisphere Publishing Corp., 1991).

World Bank Meeting in Bangkok

David Korten, *When Corporations Rule the World* (West Hartford, Connecticut: Kumarian Press, Inc., and San Francisco: Berrett-Koehler Publishers, Inc., 1995).

Loss of IQ in Bangkok's Children

United Nations Development Programme, *Human Development Report 1998* (New York and Oxford: Oxford University Press, 1998).

Chao Phraya Delta

Carmen Revenga, Sibhan Murray, Janet Abramovitz, and Allen Hammond, *Watersheds of the World: Ecological Value and Vulnerability* (Washington, DC: World Resources Institute and Worldwatch Institute, 1998).

Child Prostitutes in Thailand

Aaron Sachs, "The Last Commodity: Child Prostitution in the Developing World," *World Watch*, July/August 1994.

Consumption of Fish vs. Beef

Lester R. Brown and Michael Strauss, *Vital Signs 1998* (New York: W.W. Norton, 1998).

"Trash" Fish

International Center for Living Aquatic Resources Management, Philippines, and University of British Columbia, Canada, reported by Anne Platt McGinn in "Freefall in Global Fish Stocks," *World Watch*, May/June 1998.

UV Killing Phytoplankton

Donat P. Hader, Robert C. Worrest, H.D. Kumar, and Raymond C. Smith, "Effects of Increased Solar Ultraviolet Radiation on Aquatic Ecosystems," *Ambio*, May 3, 1995; Anne Platt McGinn, *Rocking the Boat: Conserving Fisheries and Protecting Jobs*, Worldwatch Paper 142 (Washington, DC: Worldwatch Institute, 1998).

Thai Shrimp Mafia

Anne Platt McGinn, "Blue Revolution: The Promises and Pitfalls of Fish Farming," *World Watch*, March/April 1998.

5 Million Tons of Fish Ground Up

McGinn, "Blue Revolution."

Near the Bottom in Math and Science Tests

National Center for Education Statistics, reported by Rene Sanchez in "US High School Seniors Near Bottom," *The Washington Post*, February 25, 1998.

Dreaming of Wealth

Juliet Schor, *The Overspent American: Upscaling, Downshifting, and The New Consumer* (New York: Basic Books, 1998).

The Richest 1 Percent

Paul Hawken, *The Ecology of Commerce* (New York: HarperCollins, 1993).

"We Are Poking It with Sticks"

William K. Stevens, "If Climate Changes, It May Change Quickly," *The New York Times*, January 27, 1998.

Hottest Six Months

National Oceanic and Atmospheric Administration, National

Climatic Data Center, Ashville, North Carolina http://www.ncdc.noaa.gov/ol/climate/research/1998/jul/jul98.html viewed August 10, 1998.

India's Poverty, Despite Fast-Growing GDP

Payal Sampat, "What Does India Want?" *World Watch*, September/October 1998.

Psychiatric Causes of Disability

Abigail Trafford, "World Health by the Numbers," *The Washington Post*, June 3, 1997.

Indian Farmer Suicides

Jonathan Karp, "Deadly Crop: Difficult Times Drive India's Cotton Farmers to Desperate Actions," *The Wall Street Journal*, February 18, 1998.

Chinese Farmer Suicides

Lijia MacLeod, "The Dying Fields," *Far Eastern Economic Review*, April 23, 1998.

Deltas Drying Up

Sandra Postel, "Where Have All the Rivers Gone?" *World Watch*, May/June 1995; Lester R. Brown, "China's Water Shortage Could Shake World Food Security," *World Watch*, July/August 1998.

22 Countries Short of Water

Postel, *Last Oasis*.

Desertification Affecting Over 100 countries

"Status of Desertification and Implementation of the United Nations Plan to Combat Desertification," Report of the Executive Director to the Third Special Session of the Governing Council, United Nations Environment Programme (UNEP), Nairobi, 1992.

CHAPTER 4

(NOTES FOR PAGES 125–147)

Expansions and Contractions of Civilizations

John B. Sparks, *Histomap of World History* (New York: Rand McNally, 1990).

Vanished Civilizations

Clive Ponting, *A Green History of the World: The Environment and the Collapse of Great Civilizations* (New York and London: Penguin Books, 1991; Jonathan Norton Leonard, *The First Farmers* (New York: Time-Life Books, 1973); A.H.M. Jones, *The Later Roman Empire* (Baltimore: Johns Hopkins University Press, 1964). Also see notes Pollen Analysis on Easter Island and Decline of Easter Island Civilization (below).

A Tale of Two Treaties: Montreal vs. Kyoto

Christopher Flavin, "Last Tango in Buenos Aires," *World Watch*, November/December 1998.

Ozone Still Thinning

Rumen Bojkor, World Meteorological Organization, Geneva (news release), summer 1998.

Nuclear, Biological Warfare, and CFC Threats

Ed Ayres, "The Expanding Shadow Economy," *World Watch*, July/August 1996.

Pollen Analysis on Easter Island

J.R. Flenley, "Stratigraphic Evidence of Environmental Change on Easter Island," *Asia Perspectives*, vol. 22, 1979; J.R. Flenley and S.M. King, "Late Quaternary Pollen Records from Easter Island," *Journal of Quaternary Science* 6, 1984.

New and Resurgent Diseases

The Harvard Working Group on New and Resurgent Diseases, "New and Resurgent Diseases: The Failure of Attempted Eradication," *The Ecologist,* January/February 1995; Pierro Olliaro, Jacqueline Cattani, and Dyann Wirth, "Malaria and the Submerged Disease," *Journal of the American Medical Association,* January 17, 1996; J.W. Mason, "The Plague Years," *In These Times,* March 20, 1995; Anne E. Platt, *Infecting Ourselves: How Environmental and Social Disruptions Trigger Disease,* Worldwatch Paper 129 (Washington, DC: Worldwatch Institute, 1996).

Decline of Easter Island Civilization

Jo Anne Van Tilburg, *Easter Island: Archeology, Ecology, and Culture* (Washington, DC: Smithsonian Institution Press, 1994).

China's 500 New Coal-Powered Power Plants

Hilary F. French, "When Foreign Investors Pay for Development," *World Watch,* May/June 1997.

Europeans Eating Their Own Seed Grain

Ponting, *A Green History of the World,* p. 104.

Topsoil Loss

Janet Abramovitz, "Putting a Value on Nature's 'Free' Services," *World Watch,* January/February 1998.

The First Towns

Ponting, *A Green History of the World.*

CHAPTER 5

(NOTES FOR PAGES 149–184)

Population Explosion Is Over?

Ben J. Wattenberg, "The Population Explosion Is Over," *The New York Times Magazine,* September 1997.

Another New York City Each Month

Population Reference Bureau, *World Population Data Sheet* (Washington, DC,1997).

UN Population Projections

United Nations Department of Economic and Social Affairs, Population Division, October 1998.

Corporate Concentration of Media

Mark Crispin Miller, "The Crushing Power of Big Publishing," *The Nation*, March 17, 1997.

American Enterprise Institute Mission

NIRA's World Directory of Think Tanks, Second Edition (Tokyo: National Institute for Research Advancement, 1996); Sharon Beder, *Global Spin: The Corporate Assault on Environmentalism* (White River Junction, Vermont: Chelsea Green Publishing Company, 1997).

"Science Has Spoken"

Arthur B. Robinson and Zachary W. Robinson, "Science Has Spoken: Global Warming Is a Myth," *The Wall Street Journal*, December 4, 1997.

"Environmentalism vs. Sorcery"

Ron Brunton, Australian Institute of Petroleum, cited in Sharon Beder, *Global Spin: The Corporate Assault on Environmentalism.*

"Recycling Is Garbage"

John Tierney, "Recycling is Garbage," *The New York Times Magazine*, June 30, 1996; Ed Ayres, "False Reports" (editorial), *World Watch*, November/December 1995.

"Apocalypse or Hot Air?"

Andrew Rowell, *Green Backlash: Global Subversion of the Environmental Movement* (London and New York: Routledge, 1996).

"Grow or Die"

"Forecast '98: Grow or Die!" *Your Company—The Magazine for Small Business Owners from "Money,"* 1998.

PR Industry Disguised as News Industry

See *PR Watch* (quarterly), Center for Media & Democracy, Madison, Wisconsin.

Who Owns AP?

Michael Parenti, "The Myth of the Liberal Media," *The Humanist*, January/February 1995.

"Petroleum Broadcasting System" (PBS)

Parenti, "The Myth of the Liberal Media."

Shell Oil and the Hanging of Ken Saro-Wiwa

Aaron Sachs, "Dying for Oil," *World Watch*, May/June 1996.

3,000 Human Cultures Extinguished

Estimate by Michael Krauss, Alaska Native Language Center, University of Alaska, 1992, cited in Alan Thein Durning, *Guardians of the Land: Indigenous Peoples and the Health of the Earth*, Worldwatch Paper 112 (Washington, DC: Worldwatch Institute 1992).

Extinctions of Human Cultures in the Twentieth Century (footnote)

Elaine Briere and Dan Devaney, "East Timor: The Slaughter of a Tribal Nation," *Canadian Dimension*, October 1990.

Average Child Sees 40,000 Murders

Derrick Jensen, "Telling Stories: How Television Skews Our View of Society, And Ourselves" (interview with George Gerbner), *The Sun* magazine, August 1998.

Average American Sees 150,000 Commercial Advertisements

United Nations Development Programme (UNDP), *Human Development Report 1998* (New York and Oxford, Oxford University Press, 1998).

"Pissing in the Arafura Sea"

Australian Finance Review, quoted in Curtis Runyan, "Indonesia's Discontent," *World Watch*, May/June 1998.

"Alabama Is Gonna Get Its Pants Beaten Off"

Mobile Register, quoted in Ashley T. Mattoon, "Paper Forests," *World Watch*, March/April 1998.

Eradicating Macaques

Yomiuri Shimbun, Tokyo, various issues, 1990-95.

Corporate PR

See, for example, *PR Watch: Public Interest Reporting on the PR/Public Affairs Industry*, Volume 4, Number 1, first quarter 1997.

E. Bruce Harrison

John Stauber and Sheldon Rampton, "Public Relations and Private Interests," *Earth Island Journal*, winter 1995-96.

Exxon Shareholders' Magazine

Bill Corporon, "Product of Politics, Kyoto Pact Sidesteps Science, Economics," *The Lamp*, summer 1998.

"Advancement of Sound Science" News Release

"500 Physicians, Scientists Oppose Climate Treaty," The Advancement of Sound Science Coalition, Washinton, DC, December 3, 1997.

Competitive Enterprise Institute News Release

"White House Fabricates Greenhouse Consensus: Institute Runs Ad Challenging Administration's Stance on Warming," Competitive Enterprise Institute, Washington, DC, October 4, 1997.

"American Council on Science and Health" News Release

The release included a cover letter from the council and a statement by its president, Dr. Elizabeth M. Whelan, titled "Global Warming Will Not Devastate Human Health."

Article that Looked Like a Reprint from a Scientific Journal

Arthur B. Robinson, Sallie L. Baliunas, Willie Soon, and Zachary W. Robinson, Oregon Institute of Science and Medicine, Cave Junction, Oregon, January 1998.

Wall Street Journal's "Global Warming Is a Myth" Hoax (footnote)

Letter from David E. Kennell, Ph.D., professor emeritus, Washington University School of Medicine, St. Louis, Missouri, March 24, 1998.

2,500 Economists

"The Economists' Statement on Climate Change," Redefining Progress, San Francisco, California, 1997. The original signers were Kenneth Arrow, Stanford University; Dale Jorgenson, Harvard University; Paul Krugman, MIT; William Nordhaus, Yale University; and Robert Solow, MIT. The number of signers as of March 1, 1997 was 2,509, including eight Nobel Prize winners in economics.

CHAPTER 6

(NOTES FOR PAGES 185–223)

Cesium-137

"Report on Nuclear Program Views Environmental Effects," *O Globo* (Rio de Janeiro), September 30, 1990, translated in *FBIS Daily Report/Latin America*, Rosslyn, Virginia; Gail Daneker and Jennifer Scarlott, "Nuclear Tragedy Strikes Brazil," *RWC Waste Paper* (Radioactive Waste Campaign, New York), winter 1987/1988; Greenpeace International.

A Convergence among Religious Leaders and Scientists

See, for example, Simon Oxley, "People, Land, and God," in *Contact,* a publication of the World Council of Churches, August/September 1997.

Wired to Believe

Peter Schwartz and Peter Leyden, "The Long Boom: A History of the Future, 1980-2020," *Wired*, July 1997.

Silicon Valley Millionaires

Aaron Sachs, "Virtual Ecology: A Brief Environmental History of Silicon Valley," *World Watch*, January/February 1999.

The Growing Divide between Rich and Poor

United Nations Research Institute for Social Development, *States of Disarray: The Social Effects of Globalization* (London: UNRISD, 1995); Tom Athanasiou, *Divided Planet: The Ecology of Rich and Poor* (Boston and New York: Little Brown, 1996); United Nations Development Programme (UNDP), *Human Development Report 1998* (New York and Oxford: Oxford University Press, 1998).

The Main Problem with TV

George Gerbner, Annenberg School of Communications, University of Pennsylvania, quoted in *The Sun* magazine, August 1998.

Hours of TV

Nielson Media Research, reported in *The Washington Post*, September 16, 1994.

The Overlooked $33 Trillion

Robert Costanza et al., "The Value of the World's Ecosystem Services and Natural Capital," *Nature*, May 15, 1997.

Costs of Industrial Activity Not Incorporated into Markets

Paul Hawken, *The Ecology of Commerce* (New York: HarperCollins, 1993); Ernst Ulrich von Weizacker, "Let Prices Tell the Ecological Truth," *Our Planet*, vol. 7, no. 1 (1995); David Malin Roodman, *The Natural Wealth of Nations* (New York: W.W. Norton, 1998).

7 Pounds of Grain to Produce 1 Pound of Beef

Brian Halweil (news release), Worldwatch Institute, summer 1998.

SUVs Gorging on Fuel

Molly O'Meara, "One Step Forward, Two Steps Back" (editorial), *World Watch*, May/June 1998.

Credit for Carbon Sinks

Ashley T. Mattoon, "Bogging Down in the Sinks: Escapist Accounting for Tree-Planting Schemes Don't Add Up to Climate Stability," *World Watch*, November/December 1998; David S. Schimel et al., "CO_2 in the Carbon Cycle," *Climate Change 1994* (Cambridge, UK: Cambridge University Press, 1995).

Record Number of Forest Fires

World Wide Fund for Nature (WWF), cited by Curtis Runyan in "Environmental Intelligence," *World Watch*, March/April 1998.

Most Forest Fires Set by Rich Plantation Owners

Ed Ayres, "Slash and Burn" (back cover), *World Watch*, January/February 1998; Janet Abramovitz, Worldwatch Institute, private communication, 1998.

Trees That Migrate Too Slowly

Susan L. Bassow and Peter C. Frumhoff, "The Greenhouse Challenge," *Defenders*, summer 1997.

Introduction of a New Species Can Be Like a Bomb

Chris Bright, *Life Out of Bounds: Bioinvasions in a Borderless World* (New York: W.W. Norton, 1998).

The Price on a Wild Animal's Head

Donovan Webster, "Inside the $10 Billion Black Market in Endangered Animals," *The New York Times Magazine*, February 16, 1997.

CHAPTER 7

(NOTES FOR PAGES 225–252)

Power Shifting from Nations

David Korten, *When Corporations Rule the World* (West Hartford, Connecticut: Kumarian Press, 1995); Robert Muller, "Globalization and the Fate of the Nation-State," report of the International Roundtable of the Gorbachev Foundation, the Club of Budapest and the Grauso Foundation, on Globalization and the Future of the Nation-State, Cagliari, Sardinia, May 1998.

Collapse of Grain Production in the Former Soviet Union

Brian Halweil, "Grain Production in the Former Soviet Union, Russia, and the Ukraine, 1960-1997" (unpublished), Worldwatch Institute, 1998, and UN Food and Agriculture Organization (FAO), various years.

Teenagers Driving Their Sport Utility Vehicles to the Mall

For data on SUV energy efficiency and pollution, see Friends of the Earth Roadhog Reduction Campaign at www.suv.org or the Sierra Club's Global Warming website at www.toowarm.org.

Aral Sea Ships Lying Stranded on Sand, Miles from Shore

"Down at the Mouth" (back cover), *World Watch*, May/June 1995

China's Fen River No Longer Exists

Lester R. Brown and Brian Halweil, "China's Water Shortage Could Shake World Food Security," *World Watch*, July/August 1998.

Articles Triggering NIC Response

Lester R. Brown, "Who Will Feed China?" *World Watch*, September/October 1994; Brown, "Facing Food Scarcity," *World Watch*, November/December 1995.

"National Security"

Michael Renner, *Fighting for Survival: Environmental Decline, Social Conflict, and the New Age of Insecurity* (New York: W.W. Norton & Company, 1996).

CIA Study of China

Lawrence K. Gershwin, National Intelligence Officer for Science and Technology, and Terrance J. Flannery, Director, DCI Environmental Center, Central Intelligence Agency, Memo from the National Intelligence Council on "A MEDEA Study on The Future of Chinese Agriculture," January 16, 1998.

1,000 Tons of Water, 1 Ton of Grain

Lester R. Brown, "Facing Food Scarcity," *World Watch*, November/December 1995.

Street Riots in Indonesia

Marcus W. Brauchli, "Indonesia's Struggles Become Elemental: A Shortage of Rice," *Wall Street Journal*, October 20, 1998.

You Too Can Be Rich

Eric Brown, "Spending Like There's No Tomorrow: Goodbye Joneses, Hello Bill Gates," *Enough!*, a quarterly report on consumption, quality of life, and the environment, Center for a New American Dream, fall 1998.

Poor in Rich Nations

United Nations Development Programme (UNDP), *Human Development Report 1998* (New York and Oxford: Oxford University Press, 1998).

Turning to Desert

Independent Commission on International Humanitarian Issues, *The Encroaching Desert: The Consequences of Human Failure* (London, UK and Atlantic Highlands, New Jersey: Zed Books, 1986); United Nations Convention to Combat Desertification, *New Opportunities for Development: The Desertification* Convention (Washington, DC: The World Bank, 1998).

Colorado River Fish Extinctions

Frank Wilson, "A Fish Out of Water: A Proposal for International Instream Flow Rights in the Lower Colorado River," *Colorado Journal of Environmental Law and Policy*, volume 5.249, 1994.

Water's Market Value: $200 vs. $14,000

Lester R. Brown, "China's Water Shortage Could Shake World Food Security," *World Watch*, July/August 1998.

CHAPTER 8

(NOTES FOR PAGES 253–279)

56 Million Flooded out in Yangtze Basin

Official Chinese government reports quoted in John Pomfret, "Chinese Flood Area Awash in Disorder," *The Washington Post*, August 12, 1998.

21 Million Flooded out in Bangladesh

Ed Ayres, "NOAA'S ARC," *World Watch*, November/December 1998.

Water Is the Largest Carrier of Diseases

J. M. Hunter et al., editors, *Parasitic Diseases in Water Resources Development: The Need for Intersectoral Negotiations* (Geneva: World Health Organization, 1993); Anne Platt, "Water-Borne Killers," *World Watch*, March/April 1996.

Insurance Companies Already Know the Risks

Ake Munkhammar of Skandia insurance company, H.R. Kaufman of Swiss Re insurance company, and Frank Nutter of Reinsurance Association of America, quoted in Christopher Flavin, "Storm Warnings: Climate Change Hits the Insurance Industry," *World Watch*, November/December 1994; Jessica Matthews, "If a Hurricane Hits Miami," *The Washington Post*, October 22, 1996; Andrew Marshall, Reuters, December 29, 1997: "Munich Re, the largest reinsurer, said . . . catastrophes will become more frequent and more costly. . . . Global warming, attributed to increased greenhouse gas emissions . . . poses an ever greater threat."

Vulnerability to Coastal Storm Surges

Michael Renner, *Fighting for Survival: Environmental Decline, Social Conflict, and the New Age of Insecurity.* (New York: W. W. Norton, 1996).

Leaking Genes (footnote)

Jane Rissler, senior staff scientist, Union of Concerned Scientists, quoted by Rick Weiss in "Strategy Worries Crop Up in Biotechnology's War on Pests," *The Washington Post,* September 21, 1998.

French Farmers Upset with Agribusiness (footnote)

Brian Halweil, "Bio-serfdom and the New Feudalism," *World Watch,* May/June 1998.

Genes from a Moth or an Eggplant

Partial list of engineered food organisms approved for field testing or commercialization in the United States, compiled by the Union of Concerned Scientists, Cambridge, Massachusetts, summer 1998.

"Poking It with a Stick"

Wallace S. Broecker, Lamont-Doherty Earth Observatory, Columbia University, New York, quoted in "If Climate Changes, It May Change Quickly," *The New York Times,* January 27, 1998.

Fallacy of Assuming More Consumption Means More Satisfaction

Alan Thein Durning, *How Much is Enough? The Consumer Society and the Future of the Earth* (New York: W. W. Norton , 1992).

Higher Income Leading to Higher Meat and Water Consumption

Lester R. Brown, "China's Water Shortage Could Shake World Food Security," *World Watch,* July/August 1998.

Car-Free Neighborhoods

Jane Holtz Kay, *Asphalt Nation: How the Automobile Took Over America and How We Can Take It Back* (New York: Crown Publishers, 1997).

Compact Cities

Molly O'Meara, "How Mid-Sized Cities Can Avoid Strangulation," *World Watch*, September/October 1998; Jane Jacobs, *The Death and Life of Great American Cities* (New York: Random House, 1961 and 1989).

Curitiba's Integrated Transportation System

Jonas Rabinovitz and Josef Leitmann, *Environmental Innovation and Management in Curitiba, Brazil*, UMP Working Paper Series 1 (New York: United Nations Development Programme, 1993).

Graduation Pledge of Social and Environmental Responsibility

Neil Wollman, Department of Psychology, Manchester College, North Manchester, Indiana (information sheet).

Cows, Pigs, Chickens, Etc. Produce 130 Times as Much Waste as People Do

Brian Halweil, "On July 4, United States Leads World Meat Stampede" (news release), Worldwatch Institute, Washington, DC, July 2, 1998.

10,000 Days' Energy Spent in 1 Day

Paul Hawken, *The Ecology of Commerce.*

Connections between Deforestation, Global Warming, and Flooding

Ed Ayres, "As Temperature Rises, So Does Water," *World Watch*, November/December 1998.

CHAPTER 9

(NOTES FOR PAGES 281–308)

Robert Shapiro's Epiphany

Joan Magretta, "Growth through Global Sustainability," an interview with Monsanto's CEO, Robert Shapiro, *Harvard Business Review*, January/February 1997.

Monsanto and Roundup

"An Open Letter to Robert Shapiro, CEO, Monsanto" (editorial), *The Ecologist*, September/October 1998

World Bank's Admission (footnote)

Marcus W. Brauchli, "Speak No Evil: Why the World Bank Failed to Anticipate Indonesia's Deep Crisis: It Often Soft-Pedaled Effects of Country's Corruption, Misread Extent of Poverty," *The Wall Street Journal*, July 14, 1998. The article states, "The Bank went along with government estimates that showed epic improvements in living standards, despite indications the numbers were inflated. A majority of Indonesians long were clustered only a shade above the international $1-a-day poverty standard. Now a majority are well below it."

Private Armies

Elizabeth Rubin, "An Army of One's Own," *Harper's Magazine*, February 1997; Shawshe Tourour, executive assistant to the Secretary-General, United Nations, on National Public Radio's "Morning Edition," June 10, 1997.

"Privatization Accompanied by an Upswing of Corruption"

Ildiko Ekes, Hungarian Research Institute for Economic and Social Affairs, quoted in Ed Ayres, "The Expanding Shadow Economy," *World Watch*, July/August 1996.

Economic Immigrants, Political Refugees, The Jobless, The Hungry, The Lost. . . .

Hal Kane, "What's Driving Migration," *World Watch*, January/February 1995.

Upscaling of the American Dream

Juliet Schor, *The Overspent American.*

The Border between You and Your Sociobiological World

Edward O. Wilson, *Biophilia* (Cambridge, Massachusetts: Harvard University Press, 1984).

A Matter of Scale

Ed Ayres, "Fighting Brushfires," *World Watch*, July/August 1997; Tom Athanasiou, letter to the editor, *World Watch*, September/October 1995.

Other Cultures, Species, and Generations

Jay B. McDaniel, "Emerging Options in Ecological Christianity: The New Story, the Biblical Story, and Panentheism," in *Ecological Prospects: Scientific, Religious, and Aesthetic Perspectives*, Christopher Key Chappl, editor (Albany: State University of New York, 1994).

The Copernican Insight

Jesse H. Ausubel, "The Liberation of the Environment," in *Technological Trajectories and the Human Environment*, Jesse H. Ausubel and H. Dale Langford, editors, National Academy of Engineering (Washington, DC: National Academy Press, 1997).

Separation of Powers between Corporate Largess and the Funding of Science

Mae-Wan Ho, *Genetic Engineering—Dream or Nightmare?* (Bath, UK: Gateway Books, 1998).

Complete Materials and Energy Accounting

Robert U. Ayres and Udo E. Simonis, editors, *Industrial Metabolism: Restructuring for Sustainable Development* (Tokyo: United

Nations University Press, 1994); Robert U. Ayres and Leslie W. Ayres, *Industrial Ecology: Closing the Materials Cycle* (Aldershott, UK: Edward Elgar, 1996).

Asking Markets to Do Things They Are Not Designed to Do

Amory Lovins, *Orion Quarterly,* quoted in *World Watch*, September/October 1997.

Vast Majority of All Species That Have Ever Lived Are Extinct

Edward O. Wilson, *The Diversity of Life* (Cambridge, Massachusetts: Harvard University Press, 1992).

INDEX